Proceedings of the International Neural Networks Society 4

Series Editors

Plamen Angelov, *School of Computing and Communications, University of Lancaster, Lancaster, UK*
Robert Kozma, *Optimization and Networks Department, University of Memphis, Memphis, USA*

The “Proceedings of the International Neural Networks Society INNS” publishes research contributions on fundamental principles and applications related to neural networks and modeling behavioral and brain processes. Topics of interest include new developments, state-of-art theories, methods and practical applications, covering all aspects of neural networks and neuromorphic technologies for (artificially; replace with anthropomorphic) intelligent (designs; replace with systems). This series covers high quality books that contribute to the full range of neural networks research, from computational neuroscience, cognitive science, behavioral and brain modeling, (add machine) learning algorithms, mathematical theories, to technological applications of systems that significantly use neural network concepts and techniques.

The series publishes monographs, contributed volumes, lecture notes, edited volumes, and conference proceedings in neural networks spanning theoretical, experimental, computational, and engineering aspects. Submissions of highly innovative cutting-edge contributions are encouraged, which extend our understanding at the forefront of science going beyond mainstream approaches. Of particular value to both the contributors and the readership are the short publication timeframe and the world-wide distribution, which enable both wide and rapid dissemination of research output in this rapidly developing research field.

Ivan Kolev Koychev
Editor

Recent Advances in Education and Research in the Information Society

Proceedings of the 17th Conference on Education and Research in the Information Society

Springer

Editor
Ivan Kolev Koychev
Faculty of Mathematics and Informatics
University of Sofia "St Kliment Ohridski"
Sofia, Bulgaria

ISSN 2661-8141 ISSN 2661-815X (electronic)
Proceedings of the International Neural Networks Society
ISBN 978-3-032-30152-9 ISBN 978-3-032-30153-6 (eBook)
https://doi.org/10.1007/978-3-032-30153-6

This Springer imprint is published by the registered company Springer Nature Switzerland AG
The registered company address is: Gewerbestrasse 11, 6330 Cham, Switzerland

Preface

This volume includes a selected paper from the Technology Track of the 17th International Conference on Education and Research in the Information Society, held 29–30 September 2025 in Plovdiv, Bulgaria.

The included papers propose AI-based solutions to socially significant problems, including the spread of propaganda and disinformation on social networks, improving health care, and enhancing the user experience. In particular, the papers contribute:

The first chapter by Moravski et al. proposes a new generalized annotation scheme for classifying narratives in news articles to help identify disinformation. Instead of using custom, domain-specific categories, the authors apply established literary frameworks based on emotional arcs and standard conflict categories. They created and annotated a dataset of news paragraphs using these dimensions and evaluated automatic annotation with contemporary large language models (LLMs). The results suggest that using LLMs for this type of narrative analysis is a feasible, scalable, and cost-effective approach.

This second chapter by Ribin N. et al. addresses the spread of propaganda on social media by proposing a method to detect complex persuasion techniques in text. The authors focus specifically on Bulgarian, evaluating existing language models using a dataset of news articles and parliamentary debates annotated with 23 persuasion techniques. The approach is general and can be applied to other low-resource languages.

The third chapter, authored by Vasileva, S. et al., reviews recent research on clinical entity linking, which maps clinical text to standardized medical concepts. It surveys 43 studies published between 2023 and 2025, analyzing methods, models, datasets, languages, and ontologies used. Most approaches rely on retrieve-and-rerank architectures with transformer-based models, increasingly integrating decoder models. While LLMs show promise, especially for candidate reranking and term standardization, they face challenges due to the vast number of clinical concepts and the limited availability of multilingual, labeled data. Overall, data scarcity and complex medical terminology remain key obstacles, particularly across languages.

The fourth chapter by Frantseva et al. describes a study proposing a method to distinguish Alzheimer's disease, frontotemporal dementia, and healthy individuals using resting-state EEG data. The approach combines rigorous feature selection with clustering analysis to separate the three groups, and its effectiveness is evaluated using clustering quality metrics. Results show that statistically significant EEG coherence features, particularly within subdivided beta- and alpha-frequency bands, substantially improve group discrimination. Using a semi-supervised clustering approach with well-chosen initial cluster centers further enhances classification performance.

This fifth chapter by Petrov et al. introduces "Flare," a new model that removes block artifacts and discontinuities caused by low-quality JPEG compression. Flare transforms spatial data into the frequency domain and works on both monochrome and color images of any resolution or quality. The model achieves state-of-the-art SSIM results and near-state-of-the-art PSNR, effectively reducing artifacts that rival those of existing methods.

The sixth chapter by Varlamova et al. presents a machine learning framework for real-time, behavior-based user segmentation in mobile applications. Unlike traditional methods, interpretable models maintain interpretability by employing an ensemble. Experiments show strong performance, with ROC-AUC above 0.80, demonstrating improved user prioritization, cost efficiency, and adaptive personalization in competitive mobile app environments.

Thanks to the reviewers for their thorough reviews and valuable recommendations. Thanks to the authors for submitting and preparing the final version of their research.

The open access of this volume was supported by the European Union-NextGenerationEU, through the National Recovery and Resilience Plan of the Republic of Bulgaria [Grant Project No. BG-RRP-2.004-0008].

Ivan Kolev Koychev

Contents

Generic-Level Narratives Classification Scheme

Ivo Moravski[1(✉)], Dimitar Iliyanov Dimitrov[1], Ivan Koychev[1], and Preslav Nakov[2]

[1] Faculty of Mathematics and Informatics, Sofia University "St. Kliment Ohridski", Sofia, Bulgaria
{moravski,koychev,ilijanovd}@fmi.uni-sofia.bg

[2] Mohamed bin Zayed University of Artificial Intelligence, Abu Dhabi, UAE
Preslav.Nakov@mbzuai.ac.ae

Abstract. News articles often include repeated messages pushing for a particular viewpoint on a topic. These narratives can be explicit or hidden and appear in different articles, frequently leading to the spread of disinformation. Commonly, the approaches for narratives classification rely on custom, domain-specific taxonomies. In literary theory, though, there are well established frameworks for classification of narratives at generic level. We put forward the hypothesis that applying these existing frameworks in the area of narrative classification might provide a number of benefits, e.g. classification at a generic level might be useful by itself or as a feature for training on a specialized taxonomy. To this end, we propose a Generalized Annotation Scheme. The annotation scheme classifies narratives according to established literary frameworks, specifically Booker's seven basic plot types, Reagan's six emotional arcs, and the six conflict categories. We created a dataset of news article paragraphs, based on SemEval 2025 Task 10, Subtask 2. We annotated the paragraphs in the dataset along the three dimensions of our annotation scheme. Furthermore, we explore automatic annotation conducted using recent advanced language models, both proprietary and open-source. Our results indicate that employing state-of-the-art commercial LLMs for annotation tasks is feasible, paving the way for scalable, consistent, and cost-effective narrative analysis tasks.

1 Introduction

In a world where the number of shares and likes is a stronger argument than reason and logic, the ability to (automatically) discover important, but hidden characteristics of a (news) text is an area of ever-growing interest. Such aspects might include the text's factuality, e.g., false news vs. unverified news vs. verified news. Other essential characteristics of a text under examination include different forms of bias, e.g., framing and slant. As a coherent story that connects actors and events over a timeline, the narrative is a powerful instrument of human cognition [9]. At the same time, it (inevitably) involves oversimplification, thereby exposing the cognizer to a particular cognitive bias. It replaces a

I. K. Koychev (Ed.): ERIS 2025, PINNS 4, pp. 1–18, 2026.
https://doi.org/10.1007/978-3-032-30153-6_1

complex reality with a simplified story, which, despite being coherent, could hide important contributing factors and misrepresent the proportions of other driving forces' contributions, etc. [4]. Different actors can utilize this inherent bias in any narrative to advance specific viewpoints. Thus, narratives can easily become a subtle instrument of propaganda and disinformation, even when they are not based on false news [15]. Hence, the identification and classification of narratives, particularly in news media, remain important areas of NLP research. The range of topics in the news is dynamic. The narratives therein also evolve, leading to the creation of ever-new, problem-specific taxonomies and the consequent training of different models on them. It is interesting to study whether there are more stable characteristics of the broad set of narratives, e.g., is it possible to automatically detect the presence of a narrative and its general structure? Such thinking is encouraged by literary theory, which attempts to classify the vast diversity of literary plots into a small set of categories (e.g., [20]). If we can detect narratives at a more generic level, that might give us an edge in staying on top of the news's ever-changing landscape. If it is possible to classify news articles' narratives at a generic level, it might also make it easier to specialize an optional, second-level classifier for specific taxonomies, languages, etc. Training such a generic-level classifier would require a generalized narrative classification scheme and an annotated dataset, both of which are subjects of this article. We developed a generalized annotation scheme (see 3.1) that builds upon established literary theory categorization schemes, which classifies each text along three dimensions: plot type, emotional arc, and conflict type. We further created a dataset following this schema using news articles from the SemEval 2025 Task 10, Subtask 2 dataset. We enhanced this dataset with human-generated annotations across the three dimensions of our generic-level categorization. We also developed and applied a methodology for automatic generic-level annotation and benchmarked the results against the human-defined gold labels. We present our results and conclusions regarding the feasibility of automated annotations at the generic level. Thus, the contributions of this work are the following:

- We turned to literary theory to identify classification dimensions that can provide deep insights into the narratives. We also wanted to limit the number of categories per dimension to avoid overcomplicating the annotation process. We developed a 3-dimensional generalized narratives classification scheme that meets these requirements.
- We anticipated a dataset that already categorizes narratives using a specialized taxonomy. Such a dataset would enable us to apply our generalized categories to the same data. We annotated a dataset based on the SemEval 2025 Task 10 dataset that contains news articles, already annotated with a specialized taxonomy [15]
- We created and applied an automatic annotation process that uses LLMs in two ways: with two models acting as specialized annotators and a third as a consolidator, and with a committee of models acting as annotators. We also used stand-alone models as annotators for benchmarking.

2 Related Work

The problem of narrative classification has been an object of steady interest in the NLP research community. Contemporary approaches gravitate towards employing explicit problem-specific taxonomies. The ways to develop these taxonomies vary. Some examples include expert-made taxonomies [11,18], but also "inductively"-developed taxonomies ([10]. The annotation process and the classification models, respectively, also vary. If we examine some of the taxonomies closely, we make a number of interesting observations:

- A flat taxonomy for categorizing vaccine-related narratives [11].
- A two-level taxonomy that is domain-specific, e.g., related to climate change or the Ukraine-Russia war[15] [18].
- A three-level taxonomy, developed inductively, i.e., built by experts based on observations of narratives appearing in the news (using clustering techniques) [10]. This taxonomy is also domain-specific, i.e., targeted at the COVID-19 domain.

All these taxonomies target specific problem domains and are also particular to the time frame in which the researched narratives were developed. As social and political realities evolve, so should the narratives and, hence, the taxonomies. To the best of our knowledge, current narrative classification research lacks a generic taxonomy capable of detecting narratives at the broadest levels, as suggested by literary theory. As one of this paper's contributions, we bridge this gap by introducing a generic-level annotation scheme (together with a suitable dataset and annotation methodology).

3 Dataset

3.1 Task Definition

We hypothesize that better narrative classification results can be achieved by implementing the classification in two stages, with respect to the level of generalization. The first stage is generic narrative classification: a more universal scheme grounded on literary theory. The second stage could provide specialization based on a domain-specific taxonomy. We also hypothesize that existing LLMs can perform generic-level annotation with sufficient accuracy, as such narratives are present in many texts the LLMs were exposed to during training. This would enable a faster, more efficient annotation process. As a first step towards testing these hypotheses, we need to develop a taxonomy for the generalized annotation level and set up a process for automated narrative annotation. For the first, generalized level of our two-level classifier, we sought well-established frameworks for classifying text. We have identified the following accepted categorizations:

- according to their plot [1]
- according to their emotional arc [20]
- according to conflict type [16,22]

See B for a full annotated sample (human and automatic annotations).

Generalized Categories. The first generalized level classifies narratives based on established literary frameworks, specifically, Booker's seven basic plot types [1], Reagan's six emotional arcs [20], and the six conflict categories [14,16,22].

Here's an elaboration of the generic-level classification scheme:

DIMENSION 1: PLOT ARCHETYPE

- **Overcoming the Monster:** The protagonist sets out to defeat a great evil or monstrous antagonist that threatens them or their world.
- **Rags to Riches:** The protagonist rises from humble beginnings or adversity to achieve greatness, wealth, or happiness.
- **The Quest:** The protagonist and companions set out on a journey to obtain a significant object or reach a crucial destination, facing many obstacles along the way.
- **Voyage and Return:** The protagonist travels to a strange land, faces challenges and threats, and returns home transformed by the experience.
- **Comedy:** The story is characterized by confusion, misunderstandings, or mistaken identities, ultimately leading to a happy resolution.
- **Tragedy:** The protagonist is brought to ruin or suffers extreme sorrow, often due to a personal flaw or unavoidable fate.
- **Rebirth:** The protagonist undergoes a transformation or renewal, often after a period of darkness or despair.

DIMENSION 2: EMOTIONAL ARC

- **Rise:** The emotional tone of the story steadily improves, leading to a positive or uplifting outcome.
- **Fall:** The emotional tone of the story steadily declines, leading to a negative or tragic outcome.
- **Fall-Rise:** The story begins with a decline or negative events, but ultimately recovers and ends on a positive note.
- **Rise-Fall:** The story starts positively or with hope, but ends in decline or tragedy.
- **Rise-Fall-Rise:** The emotional tone in the story rises, then falls, but ultimately recovers and ends positively.
- **Fall-Rise-Fall:** The emotional tone in the story falls, recovers, but ultimately ends in decline or tragedy.

DIMENSION 3: TYPE OF CONFLICT

- **Character vs. Self:** The protagonist faces an internal struggle, such as doubt, fear, or a moral dilemma.
- **Character vs. Character:** The protagonist conflicts with another individual or group.
- **Character vs. Society:** The protagonist challenges societal norms, laws, or institutions.
- **Character vs. Nature:** The protagonist must overcome natural forces or disasters.

- **Character vs. Technology:** The protagonist faces challenges from machines, artificial intelligence, or technological forces.
- **Character vs. Supernatural:** The protagonist confronts elements beyond the natural world, such as ghosts, fate, or the divine.

3.2 Dataset Core: News Article Paragraphs

We apply our method to form a new dataset with generalized annotations of news articles paragraphs. The news article paragraphs are extracted from the dataset of SemEval 2025 Task 10, Subtask 2. We selected a set of 35 articles containing a total of 118 paragraphs. We annotated these paragraphs along the three dimensions of the generalized categorization scheme that we propose. A sample paragraph for annotation is shown in Appendix F. An annotated paragraph is shown in Appendix B

3.3 Human-Defined Gold Labels

We created human annotations for all the 118 paragraphs of the dataset along the three dimensions of the Generalized Annotation Scheme. Each human annotation contains the selected category for each of the three dimensions - see Appendix B. (The first dimension, "Plot Type", contains 7 categories; the second dimension "Emotional Arc", contains 6 categories and the third dimension, "Conflict Type",- 6 categories.) The human annotation process involved annotation by one of the authors and consequent validation by another. These annotations were used as the gold labels for evaluating the automatic annotations (provided by different proprietary and open-source LLMs, according to the methodology).

3.4 Annotation Process

We followed (for comparison purposes) two different approaches to annotation: *Two Models, One Consolidator*, and *Committee of Models.* We also tested the models in standalone-annotation mode, also for comparison purposes.

We've employed several contemporary proprietary and open source models. Proprietary models include GPT-4.1, Claude Sonnet 4, Gemini Flash 2.5, and others. Open source models include Bggpt, LLama 4, Mistral Saba, and others. For a complete list, see A.

Two Models, One Consolidator. The process works with a pair of annotating models and a consolidator. We tested and evaluated several combinations of models. The results are summarized in a later section. We chose two LLM-based virtual annotators. They use the same prompt and send it to different LLMs (see Appendix C for the exact prompt). If the two models suggest identical annotations for the three dimensions, these consensus annotations are accepted. If they differ on any category, the same models receive a prompt that invites them to provide arguments for the classification. Each model offers arguments for their own classification. See Appendix D for the exact prompt. Both arguments are presented to an LLM that acts as a consolidator, for which we use Claude Opus 4 as a leading commercial model that utilizes "thinking" tokens. The consolidator receives the annotation guidelines and the arguments of both annotators. It then selects the better output/classification. See Appendix E for the exact prompt, and Fig. 1 for visual representation of the process.

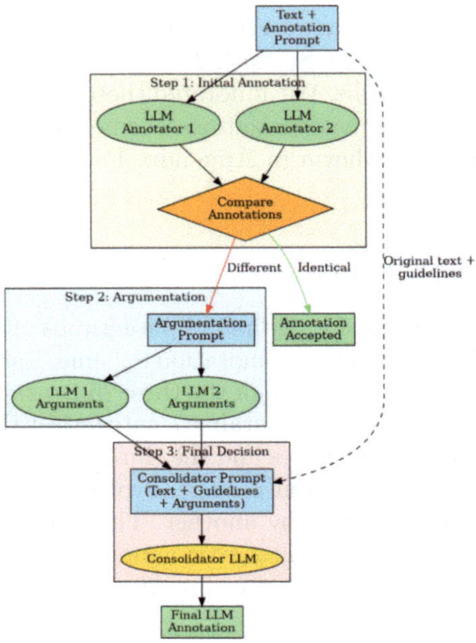

Fig. 1. *Two models, one consolidator* annotation process.

Committee of Models. The process works with a set of models. The sets are of arbitrary size. We tested and evaluated several different combinations (results are presented in 4.3). We chose several LLM-based virtual annotators. They use the same prompt and send it to different LLMs. See Appendix C for the exact prompt. The categorization of each model for each dimension in the general categorization is counted as one vote for this categorization value. A simple majority chooses the final categorization. In the event of a tie, one of the votes from the different models is selected at random. Figure 2 illustrates the committee annotation process.

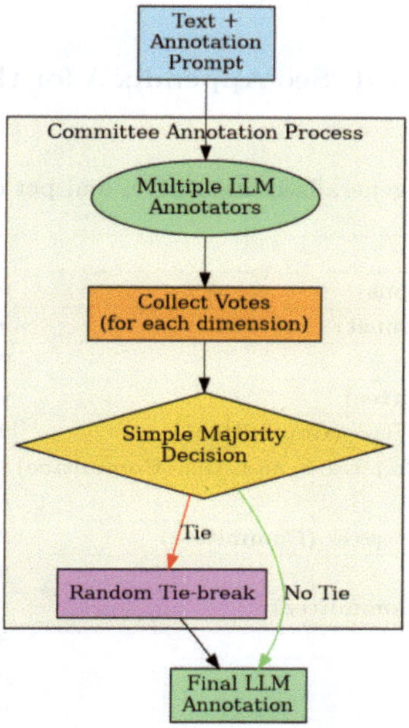

Fig. 2. Committee annotation process.

4 Experiments

4.1 Goals

The goal of the experiments is to test the accuracy of automated annotations across the three different categories of the generalized level annotation (plot, emotional arc, conflict type). The annotations are created following two distinct processes, as described in 3.4. The goal of the experiments is to test the accuracy of the automated annotations across the three different categories of the generalized level annotation (plot, emotional arc, conflict type, 3.1). The annotations are created following two different processes, as described in 3.4.

4.2 Experimental Design

We used multiple LLMs, both proprietary and open-source, as annotators for paragraphs from news articles in our dataset, following our methodology (standalone, consolidator, and committee annotation processes 1, 2). Proprietary models include GPT-4.1, Claude Sonnet 4, Gemini Flash 2.5, and others. Open source models include Bggpt, LLama 4, Mistral Saba, and others. For a complete list, see Appendix A. We benchmarked the different results against the Human-Defined Gold Labels.

4.3 Results

Annotations Leaderboard. See Appendix A for the exact names of the models.

Table 1. Accuracy per generalized annotation and per dimensional categories.

Annotator(s)	%Correct	%Arc	%Plot	%Conflict
Gpt, Sonnet, **(consolidator: Opus)**	95.20	96.61	94.92	94.07
Gpt, Sonnet, **(consolidator: Sonnet)**	94.63	97.46	93.22	93.22
Sonnet	92.94	94.07	92.37	92.37
Deepseek, Sonnet, Gpt **(Committee)**	92.66	92.37	92.37	93.22
Deepseek, Sonnet, Gpt, Gemini Flash **(Committee)**	92.09	94.07	91.53	90.68
Bggpt, Llama, Mistral, Qwen, Deepseek, Sonnet, Gpt **(Committee)**	89.83	91.53	86.44	91.53
Gpt	89.55	90.68	85.59	92.37
Bggpt, Llama, Mistral, Qwen, Deepseek **(Committee)**	89.55	92.37	85.59	90.68
Gemini Flash	87.85	93.22	88.14	82.20
Bggpt, Llama, Mistral, Qwen **(Committee)**	87.85	90.68	83.05	89.83
Deepseek	87.57	85.59	86.44	90.68
Bggpt	86.44	88.14	80.51	90.68
Mistral	85.31	85.59	80.51	89.83
Gemma3	85.03	88.98	88.14	77.97
Gpt *(Separate annotations technique)*	82.77	88.14	77.12	83.05
Sonnet *(Separate annotations technique)*	82.77	91.53	77.12	79.66
Llama	81.07	75.42	82.20	85.59
Opus	81.07	86.44	72.88	83.90
Qwen	79.10	83.05	77.12	77.12
Mercury	74.58	72.03	66.95	84.75
Gemma2	53.95	52.54	22.88	86.44

Best Identified Pair of Models. The best identified pair of models (Table 1) that performs better than all other pairs and all committees is OpenAI GPT-4.1 and Claude Sonnet 4 as annotators, and Claude Opus 4 as consolidator. They perform at 95.20% accuracy (96.61% for arc, 94.92% for plot and 94.07% for conflict dimension). Their annotations differ from humans' in less than 1 out of 20 cases. This makes the proprietary models (in a consolidator-annotation scenario) a strong tool for speeding up and streamlining the annotation process. A human annotator could review the annotations prepared by the pair of models, instead of annotating from scratch (and only occasionally make corrections).

Best Standalone Performing Model: Sonnet. Claude Sonnet 4 is the best standalone performing model (Table 1). Overall accuracy is 92.94% (94.07% for arc, 92.37% for plot and 92.37% for confilict dimension). While performing slightly worse than the winning tandem with OpenAI GPT, it outperforms the best open-source committee results.

Best Open Source Committee Annotation. The best committee annotators (Table 1) are the committee of Bggpt, Llama, Mistral, Qwen, and DeepSeek. This is, in fact, the broadest committee. Overall accuracy is 89.55% (92.37% for arc, 85.59% for plot and 90.68% for conflict dimension). It outperforms committees consisting of a subset of its members. It gets the annotations right in about 9 out of 10 cases. The best-performing proprietary pair of models gets it right in about 19 out of 20 cases. While the performance difference is significant, open-source models still seem a viable alternative to support the human annotator's labor.

Varying Temperature of a Standalone Model. We experimented with varying temperatures for the Bggpt model (Table 2). The results show that temperature can be used to squeeze additional performance out of a model.

Table 2. Varying temperature: accuracy per generalized annotation and per dimensional categories.

Annotator(s)	%Correct	% Arc	% Plot	%Conflict
Bggpt (T 0.2)	86.44	87.29	79.66	92.37
Bggpt (T 0.1)	86.44	88.14	80.51	90.68
Bggpt (T 0.4)	85.03	85.59	78.81	90.68
Bggpt (T 0.5)	84.75	85.59	80.51	88.14
Bggpt (T 0.3)	83.90	83.90	77.12	90.68
Bggpt (T 0.6)	80.79	77.97	74.58	89.83

Using the Same Model at Different Temperatures as Two Separate Annotators. We experimented with using the same model at different temperatures as two separate annotators (Table 3). The results show that the same model at various temperatures, especially when paired with a capable consolidator, can match the performance of a much stronger model.

Table 3. Using the same model at different temperatures as two separate annotators: accuracy.

Annotator(s)	Consolidator	%Correct	% Arc	% Plot	%Conflict
Deepseek	-	87.57	85.59	86.44	90.68
Bggpt (T 0.1), Bggpt (T 0.2)	Opus	87.57	88.98	81.36	92.37
Bggpt (T 0.1), Bggpt (T 0.3)	Opus	86.44	88.14	80.51	90.68
Bggpt (T 0.1), Bggpt (T 0.2)	Bggpt (T 0.1)	86.44	88.14	80.51	90.68

Annotating Separately Along the Different Dimensions. We experimented with annotating separately along the different dimensions. We provided separate prompts for each dimension. The prompts were enriched with a short description of the subcategories of each dimension. As can be seen in the scoreboard, however, the models lose accuracy when applied with this technique. A single prompt that requests categorizations across all three dimensions seems to yield superior results compared to separately categorizing along the different dimensions with three different prompts.

Other Notes

- The *thinking* versions of models perform weaker as annotators and better as consolidators.
- Deepseek's performance is strong. It is the best of the open-source models, and right after Claude Sonnet 4, GPT-4.1, and Gemini Flash 2.5. At about **87.57%** of correct annotations, it could be used as a strong helper in annotating.
- The performance of the only commercially available diffusion model, Inception Mercury, is relatively weak, compared to both proprietary and open source decoders. At **74.58%** accuracy, it is not ready for real-world annotation.
- Varying the temperature can be used for squeezing an extra percent accuracy, especially if pairing models and utilizing a capable model as a consolidator.
- The newer version of Gemma, e.g. Gemma3 outperforms Gemma2 (as expected).

Additional Observations. When reviewing the texts that the best models annotated incorrectly, we made the following additional observations:

- **"Plot type" dimension**: Irony and sarcasm are tough. It seems that identifying the Comedy plot is hardest for the LLMS. Detecting irony is a well-known problem, and we need strategies to address it.
- **"Emotional arc" dimension**: Emotional arc is the dimension that is easiest for the LLMs to categorize.
- **"Conflict type"** dimension: Distinguishing between *Character vs Character* and *Character vs Society* is tough. The differentiation is difficult, especially when characters are not individuals. Still, larger societal entities - e.g., Ukraine vs. Russia is *Character vs Character*, while, e.g., a particular committee that criticizes the state would be classified as *Character vs. Society.*

5 Conclusion and Future Work

We proposed a three-dimensional, generic narratives classification scheme. The scheme aligns with literary theory and strives to balance comprehensiveness (e.g., categorization across three dimensions) and simplicity (only 6-7 subcategories per dimension). We created a new dataset with generalized annotations see (B. We also proposed two LLM-based automatic annotation processes. The first process involves two models as annotators and a third model as a consolidator. The second consists of a committee of models for annotation. We experimented with a single prompt for annotation across all three dimensions vs separate prompts for each dimension. We achieved better results with the former approach. Overall, the results for using LLMs as annotators on the generic level of our classifier seem promising. Depending on computational and budget constraints, one could use a team of two strong models as annotators and a third strong model as a consolidator for the best results, or use an open-source committee that can still provide a strong annotation basis for human review. Using a single strong model is also a viable option. In addition to the best proprietary models, a strong open-source model, e.g., DeepSeek, can also be used. Both annotation processes yield good results and appear feasible for real-world use.

Some possible directions for future work include:

- Improve the annotation methodology. Make further experiments with the prompts, e.g. provide more examples in the prompt (few shots prompting).
- Experiment with different algorithms for tie-up resolution in committee annotation. e.g., use a consolidator, or pick the annotation by the model with better overall scores, or pick the model with better scores for a particular category, etc.
- Extend the list of models with other recent open-source models, e.g., Kimi K2, etc.
- Further experiment with varying temperature to squeeze additional performance out of the best models.
- Experiment with using different models for the different dimensions
- We should consider strategies for better handling irony and sarcasm (thereby enabling better categorization along the plot type dimension, especially for the values tragedy and comedy).
- An important next step would be to train a classifier on the annotated dataset and evaluate its performance.

Acknowledgments. This work was partially supported by the European Union-NextGenerationEU, through the National Recovery and Resilience Plan of the Republic of Bulgaria [Grant Project No. BG-RRP-2.004-0008]. Views and opinions expressed are, however, those of the author only and do not necessarily reflect those of the European Union. Neither the European Union nor the granting authority can be held responsible for them.

Declaration on Generative AI. During the preparation of this work, the author(s) used Grammarly to perform a grammar and spelling check. After using these tool(s)/ service(s), the author(s) reviewed and edited the content as needed and take(s) full responsibility for the publication's content.

Appendices

A List of Models

Table 4.

Table 4. Models that were used in the experiments.

Name	Full Name	License	Citation
Gpt	gpt-4.1	Proprietary	[17]
Sonnet	claude-sonnet-4-20250514	Proprietary	[2]
Opus	claude-opus-4-20250514	Proprietary	[2]
Gemini	gemini-2.5-flash	Proprietary	[7]
Mercury	mercury	Proprietary	[8]
Llama	llama-4-scout-17b-16e-instruct	Open Source	[12]
Deepseek	deepseek-r1-distill-llama-70b	Open Source	[3]
Qwen	qwen/3-32b	Open Source	[19]
Qwq	qwen-qwq-32b	Open Source	[21]
Gemma2	gemma-2-27b-it	Open Source	[6]
Gemma3	gemma-3-27b-it	Open Source	[5]
Mistral	mistral-saba-24b	Open Source	[13]

B Annotated Sample From Dataset


```
{
    "paragraph_id": 33,
    "text": [
        "Terrifying maps reveal the areas ",
        "most likely to be scorched ",
        "by record heat this year thanks to El NiÃśo."
],
"generalized_annotation_bggpt_0.1":
    {
        "plot": "Tragedy",
        "arc": "Fall",
        "conflict": "Character vs. Nature"
    },
...
"...committee_deepseek_claude_openai_gemini_flash":
    {
```

```
        "plot": "Tragedy",
        "arc": "Fall",
        "conflict": "Character vs. Nature"},
 ...
"generalized_annotation_gemma3":
    {
        "plot": "Tragedy",
        "arc": "Fall",
        "conflict": "Character vs. Nature"
    }
 ...
}

{"paragraph_id": 44,

    "text": [
"Speaking about the findings, ",
"UN Secretary-General AntÃşnio Guterres emphasized ",
"how quickly the world is moving ",
"in the wrong direction ",
"and away from stabilizing its climate system."
],
"generalized_annotation_human":
{"plot": "Tragedy",
"arc": "Fall",
"conflict": "Character vs. Nature"},

"generalized_annotation_bggpt_0.1":
{"plot": "Tragedy",
"arc": "Fall",
"conflict": "Character vs. Nature"},

"generalized_annotation_llama_scout":
{"plot": "Tragedy",
"arc": "Fall",
"conflict": "Character vs. Nature"},

"generalized_annotation_gemma2":
{"plot": "Tragedy",
"arc": "Fall",
"conflict": "Character vs. Nature"},

"generalized_annotation_mistral_saba":
{"plot": "Tragedy",
"arc": "Fall",
```

"conflict": "Character vs. Nature"},

"generalized_annotation_deepseek":
{"plot": "Tragedy",
"arc": "Fall",
"conflict": "Character vs. Nature"},

"generalized_annotation_qwen_qwq":
{"plot": "Tragedy",
"arc": "Fall",
"conflict": "Character vs. Nature"},

"...committee_bggpt_llama_mistral_qwen":
{"plot": "Tragedy",
"arc": "Fall",
"conflict": "Character vs. Nature"},
...
}

C LLM Prompt: Annotation Guidelines

```
You are an expert narrative annotator. Annotate the
    following Bulgarian text according to these guidelines
    :

Step 1: Generalized Narrative Annotation
- Level 1: Plot Archetype (choose one): Overcoming the
    Monster, Rags to Riches, The Quest, Voyage and Return,
     Comedy, Tragedy, Rebirth
- Level 2: Emotional Arc (choose one): Rise, Fall, Fall-
    Rise, Rise-Fall, Rise-Fall-Rise, Fall-Rise-Fall
- Level 3: Type of Conflict (choose one): Character vs.
    Self, Character vs. Character, Character vs. Society,
    Character vs. Nature, Character vs. Technology,
    Character vs. Supernatural

Examples:

Text: "Climate activists claim humanity faces an
    existential threat unless drastic measures are adopted
     immediately."
Annotations:
- Plot Archetype: Tragedy
- Emotional Arc: Fall
```

```
- Conflict: Character vs. Nature

Text: "Renewable energy advocates clash with policymakers
    who claim such technologies harm the economy."
Annotations:
- Plot Archetype: Comedy
- Emotional Arc: Rise-Fall
- Conflict: Character vs. Society

Text: "Ukraine is portrayed by Western governments and
   media as a heroic democracy fighting evil, yet reality
    shows it's deeply corrupt and a mere puppet in the
   hands of the West."
Annotations:
- Plot Archetype: Tragedy
- Emotional Arc: Fall
- Conflict: Character vs. Society

All texts will be in Bulgarian. All annotations must be
   in English.

For each classification dimension ("plot", "arc", "
   conflict"), output exactly one string value (not a
   list or array). Respond in the following JSON format,
   where each value is a single string (not a list):

{
  "generalized_annotation_llm": {
    "plot": "<one string value>",
    "arc": "<one string value>",
    "conflict": "<one string value>"
  }
}
```

D LLM Prompt: Argument

```
An expert annotator received the following annotation
   guidelines:
'''
{annotation_guidelines}.
'''

Part of the annotator's annotation for the following
   Bulgarian text was {dimension}: {value}.
```

```
Text: {text}

Please provide a concise argument (in English) justifying
    her choice for {dimension}.
```

E LLM Prompt: Consolidator

```
Two expert annotators received the following annotation
    guidelines:
‘‘‘
{annotation_guidelines}.
‘‘‘

The two expert annotators disagreed on the {dimension}
    for the following Bulgarian text:
Text: {text}

Annotator 1 chose: {openai_value}
Argument: {openai_arg}

Annotator 2 chose: {claude_value}
Argument: {claude_arg}

As a senior narrative annotator, decide which annotation
    is more appropriate.
Respond with only the winner: either 'Annotator 1' or '
    Annotator 2'. Do not quote the value itself, just the
    winner.
```

F Sample Paragraphs for Annotation (English Translation)

```
1) Shocking by Bill Gates: In 2 years, I'll be feeding
    you with tasty chemical oils.


2) "Political and industrial leaders must realize that we
     are on the brink of disaster. The science is clear,
    and the stakes for all of humanity could not be higher
    ," emphasized a spokesperson for UN Secretary-General
    AntÃşnio Guterres.
```

3) On January 7, they recorded −34ÂřC, and the (electric) buses were stopped because, you see, they can't run in such cold weather and have to be parked indoors to warm up.

References

1. Abbott, H.: The Cambridge Introduction to Narrative. 2nd edn. Cambridge University Press (2002). https://doi.org/10.1017/CBO9780511816932
2. Anthropic: System card: Claude opus 4 & claude sonnet 4. Technical Report, Anthropic (2025). https://www.anthropic.com, official system card with safety evaluations and model details
3. DeepSeek-AI et al.: DeepSeek-R1: incentivizing reasoning capability in LLMs with reinforced self-training. arXiv preprint arXiv:2501.12948 (2025)
4. Dennison, J.: Narratives: a review of concepts, determinants, effects, and uses in migration research. Comp. Migr. Stud. **9**(1), 50 (10 2021). https://doi.org/10.1186/s40878-021-00259-9
5. Gemma Team: Gemma 3 technical report. arXiv preprint arXiv:2503.19786 (2025)
6. Google: Gemma 2: improving open language models at a practical size. arXiv preprint arXiv:2408.00118 (2024)
7. Google: Gemini 2.5: pushing the frontier with advanced reasoning and multimodal capabilities. arXiv preprint arXiv:2507.06261 (2025)
8. Inception Labs et al.: Mercury: ultra-fast language models based on diffusion. arXiv preprint arXiv:2506.17298 (2025)
9. Keith Norambuena, B.F., Mitra, T., North, C.: A survey on event-based news narrative extraction. ACM Comput. Surv. **55**(14s) (2023). https://doi.org/10.1145/3584741
10. Kotseva, B., et al.: Trend analysis of COVID-19 mis/disinformation narratives-a 3-year study. PLoS ONE **18**(11), e0291423 (2023). https://doi.org/10.1371/journal.pone.0291423
11. Li, Y., Scarton, C., Song, X., Bontcheva, K.: Classifying COVID-19 vaccine narratives. In: Mitkov, R., Angelova, G. (eds.) Proceedings of the 14th International Conference on Recent Advances in Natural Language Processing, pp. 648–657. INCOMA Ltd., Shoumen, Bulgaria, Varna, Bulgaria (2023). https://aclanthology.org/2023.ranlp-1.70/
12. Meta: The Llama 4 herd: the beginning of a new era of natively multimodal AI. Meta AI Blog (2025). https://ai.meta.com/blog/llama-4-multimodal-intelligence/
13. Mistral AI Team: Mistral saba: a regional foundation model. Blog post (2025). https://mistral.ai/news/mistral-saba, 24B parameter model for Arabic and South Asian languages
14. Murtadho, A.M., Supriatna, E.: Suhendar: Analysis of daniel plainview's character traits in the movie "there will be blood" by paul thomas anderson. J. English Lang. Literat. **6**(2), 83–98 (2021). https://doi.org/10.37110/jell.v6i2.127, https://journal.stibaiec-jakarta/ojs/index.php/jell

15. Nikolaidis, N., et al.: PolyNarrative: a multilingual, multilabel, multi-domain dataset for narrative extraction from news articles. In: Che, W., Nabende, J., Shutova, E., Pilehvar, M.T. (eds.) Proceedings of the 63rd Annual Meeting of the Association for Computational Linguistics (Volume 1: Long Papers), pp. 31323–31345. ACL, Vienna, Austria (2025). https://doi.org/10.18653/v1/2025.acl-long.1513, https://aclanthology.org/2025.acl-long.1513/
16. Nur, A., Yusuf, R.: Social and internal conflict in stephenie meyer's novel "breaking dawn": anthropology study of literature. Jentera: Jurnal Kajian Sastra **12**(1), 109 (2023), konflik Internal dan Sosial dalam Novel "Breaking Dawn" Karya Stephenie Meyer: Kajian Antropologi Sastra
17. OpenAI: GPT-4 technical report. arXiv preprint arXiv:2303.08774 (2023)
18. Piskorski, J., et al.: SemEval 2025 task 10: multilingual characterization and extraction of narratives from online news. In: Rosenthal, S., Rosá, A., Ghosh, D., Zampieri, M. (eds.) Proceedings of the 19th International Workshop on Semantic Evaluation (SemEval-2025), pp. 2610–2643. ACL, Vienna, Austria (2025). https://aclanthology.org/2025.semeval-1.331/
19. Qwen Team: Qwen3 technical report. arXiv preprint arXiv:2505.09388 (2025), qwQ reasoning model discussed within this report
20. Reagan, A.J., Mitchell, L., Kiley, D., Danforth, C.M., Dodds, P.S.: The emotional arcs of stories are dominated by six basic shapes. EPJ Data Sci. **5**(1) (2016). https://doi.org/10.1140/epjds/s13688-016-0093-1
21. Team, Q.: Qwq-32b: embracing the power of reinforcement learning (2025). https://qwenlm.github.io/blog/qwq-32b/
22. Tehseem, T., Ali, I.: The structural analysis of 'thirteen reasons why' a novel by jay asher. Acad. Res. Int. **6**(1), 334 (2015). www.journals.savap.org.pk

Persuasion Technique Detection in Bulgarian

Nikolay Ribin[1(✉)], Dimitar Iliyanov Dimitrov[1], Ivan Koychev[1], and Preslav Nakov[2]

[1] Faculty of Mathematics and Informatics, Sofia University "St. Kliment Ohridski", Sofia, Bulgaria
{ribin,ilijanovd,koychev}@fmi.uni-sofia.bg
[2] Mohamed bin Zayed University of Artificial Intelligence, Abu Dhabi, UAE
Preslav.Nakov@mbzuai.ac.ae

Abstract. As social media continues gaining a significant foothold in our everyday lives, it has become a potential channel for spreading propaganda. One way to address this problem is to identify the use of persuasion techniques in text. It is a challenging task because such techniques are complex making them difficult to recognize while reading. The paper presents an approach for detecting persuasion techniques used in text. We focus our work on a Bulgarian corpus, composed of news articles and parliament debates, both accurately annotated with 23 persuasion techniques. We provide insight into existing Bulgarian language models by evaluating them on this subjective task. For this purpose, we conduct experiments involving three paradigms for a multinomial classification task: (*i*) a traditional ML model, (*ii*) a fine-tuned transformer-based model, and (*iii*) a prompt-based LLM. We use different approaches, including two levels of distinction: the classification level (paragraph or sentence) and the type of classification (binary or multiple). Our analysis of the results showed that monolingual state-of-the-art models outperform their multilingual counterparts on low-resource languages such as Bulgarian.

1 Introduction

Social and mass media encompass a wide range of communication channels that play a crucial role in shaping public opinion. They can be used to spread propaganda, which is the intentional dissemination of misleading information to promote specific agendas or ideologies, as well as to achieve a predetermined goal, as described in IPA[14]. The influence of potential propaganda can be significant, and analyzing such channels can help prevent the spread of disinformation.

Propaganda has been studied for many years and in recent years a few datasets have been developed: (*i*) TSHP-17 [20], consisting of document-level annotated articles with four classes; (*ii*) QProp [4], which uses binary annotation (propaganda vs. non-propaganda); (*iii*) PTC [7], consisting of articles, annotated on a fragment level with 18 propaganda techniques; (*iv*) The Shared

I. K. Koychev (Ed.): ERIS 2025, PINNS 4, pp. 19–30, 2026.
https://doi.org/10.1007/978-3-032-30153-6_2

Task 3 at CLEF-2024 [17], which presents a multilingual dataset covering a total of 13 languages with manually annotated 23 persuasion techniques on a fragment level; (*v*) The Shared Task on the Detection and Classification of Persuasion Techniques in Texts for Slavic Languages [18] which includes parliament debates and social media posts in five Slavic languages, annotated on a fragment level using 25 persuasion technique taxonomy.

In this paper, we aim to use a compact, high-quality, manually annotated dataset, which requires further preprocessing to be adapted for the task at hand. The dataset consists of news articles and parliament debates in Bulgarian provided by previous shared tasks on persuasion technique detection [17] [18]. The former competition on persuasion technique detection at CLEF-2024 CheckThat! focuses on developing models for detecting and classifying persuasion techniques at the span level. In contrast, the task at SlavicNLP consists of two subtasks: (*i*) binary detection of persuasion techniques in text fragments and (*ii*) multi-label classification of persuasion techniques from a given extended taxonomy derived from SemEval 2023 Task 3 [19]. To unify the data from both tasks, we are using the base taxonomy with 23 persuasion techniques.

Different approaches are being considered to evaluate the behaviour of state-of-the-art models, including multilingual transformer-based models, such as Multilingual BERT[8], Bulgarian BERT (BERT model, trained on a Bulgarian literature), and Bulgarian GPT, based on Gemma 2, introduced by Morgane Rivière et al. [21], which are well-suited for the Bulgarian language. We decompose the task into four distinct tasks with varying difficulty. The goal is to demonstrate a straightforward application of some modern models with additional fine-tuning and to share insights regarding their performance.

Our contributions can be summarised as follows:

- We collect a dataset by merging two existing datasets, consisting of news articles and parliament debates in Bulgarian, annotated with 23 labels, representing persuasion techniques.
- Based on the collected data, we propose several classification tasks, that offer different levels of complexity for more precise performance evaluation of classification models.
- We perform experiments with several state-of the-art models, which present their limits and further confirm the importance of using pretrained monolingual models on low-resource languages.

2 Related Work

2.1 Persuasion Technique Detection

Snow et al. [22] made an attempt to differentiate propaganda from persuasion and to demonstrate a separation between them according to purpose and process as well as to reveal their similarities. Nevertheless, despite their differences in techniques and goals, work on persuasion overlaps to a large extent with work on propaganda, as there is a lot of crossover between the two. Therefore, below we will discuss work on both.

Rashkin et al. [20] used their own TSHP-17 dataset on a text classification task with four defined classes: trusted, satire, hoax, and propaganda. Their approach, which uses a word n-gram language model with logistic regression, yielded promising results only for articles from sources used to train the system.

To address TSHP-17's limitation on a low number of sources considered per class, Barrón-Cedeno et al. [4] released an imbalanced dataset, QProp, with two classes: propaganda and non-propaganda. They used both TSHP-17 and QProp on a binary classification setting: detecting propaganda vs non-propaganda. They conducted a massive array of experiments, investigating various representations and trained models using logistic regression and SVMs. Their results confirmed that using distant supervision in conjunction with detailed representations might encourage the model to predict the source rather than to distinguish precisely between propaganda and non-propaganda. In another instance, Habernal et al. [11] created a more advanced corpus with more than 1000 fallacious arguments, related to ad hominem, appeal to emotion, red herring, hasty generalization, and irrelevant authority, which can be directly associated with specific propaganda techniques.

More thorough research was done by Da San Martino et al. [7]. They introduced a new dataset containing news articles annotated with 18 fine-grained propaganda techniques on a fragment level this enabled the following two tasks: (*i*) binary classification given a sentence in an article, predict whether any of the given 18 techniques has been used in it; (*ii*) multi-label multi-class classification and span detection task given a raw text, identify both the specific text fragments where a propaganda technique is being used as well as the type of the technique. Moreover, a multi-granular deep neural network is proposed to extract additional knowledge from the sentence-level task, thereby improving the fragment-level classifier. Subsequently, Da San Martino et al. [6] developed a system and made it publicly available. Most studies in the field generally used English corpora. Piskorski et al. [19] created a corpus of news articles in nine languages, enabling further research into the development of multilingual models. Other works explore propaganda and metaphor, done by Rodríguez et al. [2], propaganda and fake news, done by Huang et al. [13], and propaganda in social media, done by Nakov et al. [16].

2.2 Related Shared Tasks

Several shared tasks related to the detection of propaganda or persuasion techniques in text have been organized in the past. Semeval-2020 task 11 on Detection of Propaganda Techniques in News Articles [5] focuses on news articles - the task involves finding a text span and recognizing one of 14 defined propaganda techniques. Another closely related initiative is the NLP4IF-2019 task on Fine-Grained Propaganda Detection [7], which involves text-span detection of 18 propaganda techniques in news articles. The SemEval 2023 task 3: Detecting the Category, the Framing, and the Persuasion Techniques in Online News in a Multi-lingual Setup [19] concentrates on detecting 23 well-defined persuasion techniques in news articles covering nine languages. Both WANLP'2022 [1] and ArabicNLP'2023 [12] shared tasks included 20 propaganda technique detection in Arabic tweets and news articles. The SemEval-2021 Task 6 on Detection of Persuasion Techniques in Texts and Images [10] emphasized identifying 22 persuasion techniques in memes. Building on this prior work, the English dataset's size has been further extended to 10K memes, and three new languages were added for the SemEval-2024 Task 4: Multilingual Detection of Persuasion Techniques in Memes [9]. This task presented persuasion techniques in a newly formed hierarchy, enabling better system predictions in scenarios with low confidence in persuasion predictions. The SlavicNLP 2025 Shared Task: Detection and Classification of Persuasion Techniques in Parliamentary Debates and Social Media [18] is the most recent work, which extended the taxonomy used in previous shared tasks to 25 fine-grained persuasion techniques and covers 5 Slavic languages in total.

3 Dataset

3.1 Collection

This work uses a pre-existing datasets of Bulgarian news articles and parliamentary debates, presented and described in The Shared Task 3 at CLEF-2024 [17] and SlavicNLP 2025 Shared Task [18], respectively. The extraction of news articles is carried out using the Trafilatura tool [3], with 5–10% of cases requiring manual extraction by marking, copying, and pasting text from a relevant web page. Several unconfirmed sources were used, mainly focused on the most current topics: climate change, the Russia-Ukraine war, COVID-19, migration, and the Israeli-Palestinian conflict. The data is filtered in several stages the first involves a quick review of all articles and manually choosing articles potentially containing persuasion techniques. The news articles, except for those related to the Israeli-Palestinian conflict, were fed to a trained model for recognising propaganda in text. As a result, the model showed that the majority of the provided articles are dominated by sentences devoid of propaganda techniques. According to the trained model, the Russia-Ukraine war is the most saturated with propaganda techniques out of the four covered topics. The parliamentary debates were extracted from the official government website. Due to the large amount of text, only relevant excerpts of the debates were taken for the task.

3.2 Annotation

The annotation process is described in detail for news articles in [17] and for parliamentary debates in [18]. A team of annotators, fluent in Bulgarian, who had also performed such annotations in the past, was tasked with the annotation process; there were regular meetings to discuss complex cases with more experienced annotators who had previously participated in one or more annotation campaigns using the same taxonomy. Each document was annotated by two annotators, and one curator reconciled the annotations. The annotation was performed in two phases: in the first phase, annotators independently annotated the news articles and the parliamentary debates; in the second phase, all annotators met with a curator to discuss further and select the final gold labels.

An approach described by Piskorski et al. [19] was used to train annotators. Note that in the parliament debates, the "Appeal to pity" persuasion technique was excluded to synchronise the two datasets.

3.3 Statistics

Table 1 shows the number of annotated documents and their corresponding number of paragraphs and sentences for each subtask. The data for each subtask was split into training, validation, and test sets, as shown in the table. The label distribution for all subtasks is presented in Table 2. Overall, the most commonly annotated persuasion techniques were Loaded language, Name-calling, Casting doubt, and Questioning the Reputation. However, the specific distribution varies between the annotated dataset at the paragraph level and at the sentence level.

All annotators used the INCEpTION platform [15], a tool designed explicitly for multi-label classification tasks, which enables faster and more precise text processing.

Table 1. Dataset statistics for each *source* and *subtask*.

Source	# of documents	# of paragraphs				# of sentences			
		Train	Dev	Test	Total	Train	Dev	Test	Total
News articles	100	732	92	92	916	1,168	147	146	1,461
Parliament debates	79	507	0	1,399	1,906	1,322	0	3,893	5,215
Total	**179**	**1,239**	**92**	**1,491**	**2,822**	**2490**	**147**	**4,039**	**6,676**

4 Experiments and Evaluation

We conduct multiple experiments with a Naïve Bayes model used as a baseline and five transformer-based models across four text classification tasks to investigate the abilities of pretrained models for a specific low-resource language. An analysis is provided to evaluate the impact factor concerning the classification level.

Table 2. Dataset label distribution for each subtask.

Persuasion Technique	# of persuasive sentences	# of persuasive paragraphs
Doubt	944	516
Questioning the Reputation	829	432
Loaded language	691	578
Name-Calling Labeling	582	465
Appeal to Hypocrisy	333	164
Exaggeration-Minimisation	318	279
Appeal to Fear-Prejudice	299	214
Repetition	240	167
Conversation Killer	218	144
False Dilemma-No Choice	210	139
Causal Oversimplification	210	138
Consequential Oversimplification	191	111
Obfuscation-Vagueness-Confusion	189	139
Appeal to Values	177	134
Whataboutism	173	89
Flag Waving	168	105
Red Herring	166	78
Strawman	149	90
Appeal to Popularity	114	77
Guilt by Association	96	70
Appeal to Authority	94	61
Slogans	89	76
Appeal to Time	37	30

4.1 Experiment Design

The experiments aim to evaluate the extent to which specific models can determine whether a given Bulgarian text fragment contains any persuasion techniques. The text fragments are presented in two forms as paragraphs (can be a sequence of 1 or more sentences) or as sentences (every sentence is treated individually). We conduct all experiments by using a preprocessed variant of the merged dataset, described in Sect. 3. The preprocess includes: creating text fragments (paragraphs/sentences), extracted from the provided news articles or parliament debates; creating a vector with a length, depending on whether we use binary or multiple types of classification, for each text fragment, representing all persuasion techniques.

4.2 Models

We evaluate three paradigms for the multinomial classification task: (*i*) a traditional ML model, (*ii*) a fine-tuned transformer-based model, and (*iii*) a prompt-based LLM. As a baseline, we use a method that adapts the traditional Naive Bayes classifier to handle multi-label instances. In experiments where the setup involves binary classification, we use only two labels to indicate whether a paragraph or a sentence contains persuasion techniques. For the multi-label task, we use all possible combinations of persuasion techniques that might be used in a text. The formula can determine the total label count:

$$\sum_{i=0}^{n-1} 2^i = 2^n - 1 \tag{1}$$

For 23 possible classes, we get 8,388,607 labels. We use both monolingual and multilingual pretrained BERT-based [8] models: multilingual BERT and BG BERT with different number of parameters (BERT model trained on a Bulgarian dataset). All BERT-based models are used as encoders and are fine-tuned for the specific task. The BgGPT model is deliberately chosen to evaluate a monolingual LLM's potential for detecting persuasion techniques when provided with task insight. For this purpose, we adapt the prompt depending on the experiment, described in Sect. 4.3.

4.3 Experimental Setup

The data is split into training, development and testing, as shown in Table 1, respectively with 1,239 (44%), 92 (3%) and 1,491 (53%) examples on the paragraph level, as for the sentence level task the split is done into the following proportion of examples: 2,490 (37%), 147 (2%) and 4039 (61%). We consider two F1-score variants micro and macro F1-scores when evaluating classifier performance to offer a more comprehensive picture in a multi-class multi-label task with a highly imbalanced label distribution.

All BERT-based models are trained on a GeForce RTX-4090-24GB GPU with the following manually tuned values of the hyperparameters on the development dataset: batch size of 12 and a sequence length varying between 200 and 512, depending on the classification level of the task. For an optimiser, AdamW is used with a learning rate of 2e-5, epsilon of 1e-8, and weight decay of 0.01. One-fifth of the training steps are considered warm-up steps. All reported results are averaged across five runs with random seeds, and thresholds of 0.7 and 0.3 are used for binary and multi-class classification tasks, respectively. For tokenization, their respective tokenizers are used (multilingual and monolingual).

The BgGPT model is used in a few-shot learning setup, where a few examples are provided in the prompt. In a binary classification task, at least two examples from the training set are provided: one without a persuasion technique and one with at least one persuasion technique. In a multiple classification task, at least one instance of each persuasion technique is provided in the training set. In both cases, there is an introductory part that provides instructions to the model to act as an agent, determining whether a given paragraph or sentence contains a persuasion technique.

4.4 Results

Table 3. Results on persuasion technique detection on all subtasks. The main evaluation measure is micro-average F1-score. If there is more than one matching top micro-average F1-score, the macro-average F1-score is taken into consideration.

Classification level	Type of classification	Model	F1-micro	F1-macro
Paragraph	Binary	1. Naïve Bayes (Baseline)	0.47	0.43
		2. Multilingual BERT base (cased)	0.80	0.64
		3. BG BERT (uncased)	0.78	0.66
		4. BG BERT Large (uncased)	**0.80**	**0.68**
		5. BG BERT Extra Large (uncased)	0.80	0.59
		6. BG GPT (27b)	0.70	0.69
	Multiple	7. Naïve Bayes (Baseline)	0.01	0.01
		8. Multilingual BERT base (cased)	0.35	0.13
		9. BG BERT (uncased)	0.40	0.19
		10. BG BERT Large (uncased)	0.41	0.19
		11. BG BERT Extra Large (uncased)	**0.41**	**0.20**
		12. BG GPT (27b)	0.28	0.16
Sentence	Binary	13. Naïve Bayes (Baseline)	0.32	0.38
		14. Multilingual BERT base (cased)	0.67	0.67
		15. BG BERT (uncased)	**0.77**	**0.73**
		16. BG BERT Large (uncased)	0.74	0.62
		17. BG BERT Extra Large (uncased)	0.68	0.62
		18. BG GPT (27b)	0.64	0.63
	Multiple	19. Naïve Bayes (Baseline)	0.01	0.02
		20. Multilingual BERT base (cased)	0.24	0.08
		21. BG BERT (uncased)	0.27	0.11
		22. BG BERT Large (uncased)	0.27	0.12
		23. BG BERT Extra Large (uncased)	**0.30**	**0.16**
		24. BG GPT (27b)	0.24	0.10

The results from all experiments, described in Sect. 4.1, are shown in Table 3. In the first setting (binary classification on the paragraph level), three models (rows 2, 4, 5) achieve the best micro F1-score, whereas the best macro F1-score goes to the BG GPT model. In cases where we observe more than one best micro F1-score, we also take macro F1-score into consideration, which makes the BG BERT (Large) model the best performer. In the second setting (multiple classification on the paragraph level), BG BERT (Extra Large) outperforms BG BERT (Large) only by macro F1-score. In the third setting (binary classification on the sentence level), the best score is achieved by BG BERT (base), shown on row 15. In the fourth setting (multiple classification at the sentence level), we observe a roughly 3 times drop in micro F1-score compared to the results in the third setting.

These results show that state-of-the-art models pretrained and fine-tuned on a specific language perform better than their multilingual counterparts. They also show that LLMs are still struggling with complex tasks, even when fed additional insight. It's important to note that during the testing phase of the BG GPT, the model's answers sometimes deviated from expectations. The agent was given only two options: yes or no. In the multiple classification settings, for every paragraph or sentence, the model was prompted for each persuasion technique individually. In both binary and multiple classification settings, the model did not give a clear answer. In such cases, the answer was counted as a negative. Overall, we can conclude that monolingual approaches are necessary to achieve better results in detecting persuasion techniques in texts when a low-resource language is involved. The results show that performance drops drastically when we try to detect persuasion techniques at a lower level. Previous work by Piskorski et al. [17] shows that performance drops significantly for a span-level text classification task, indicating the need to add new features to existing transformer-based encoder models.

5 Conclusion and Future Work

We have collected a dataset of Bulgarian news articles and parliamentary debates, annotated with 23 labels, each representing a persuasion technique. We have explored the dataset by using two approaches: one annotated at the paragraph level and another at the sentence level. For each approach, we added two classification tasks: binary and multi-label. Our analysis of the two approaches reveals a degree of similarity in the distribution of persuasion techniques between datasets in other languages with the same taxonomy. For each task, we provided evidence of all models' behaviour represented by F1-micro and F1-macro scores. Our results confirmed that state-of-the-art models, pretrained on a low-resource language, tend to perform better than their multilingual counterparts.

In future work, we plan to expand the dataset by adding articles on other topics. We also plan to develop new models explicitly aimed at span-level identification of fine-grained persuasion techniques in Bulgarian texts.

Limitations

Dataset Representativeness: The datasets used in our work cover news articles from Bulgarian media and parliamentary debates, held in Bulgaria, as we strove to include a balanced representation of points of view on various topics shared by journalists, news editors and utterances of speakers covering a wide political spectrum. However, we must emphasize that the data should not be considered representative for carrying out other types of research, nor should it be considered as balanced in any way.

Biases: The datasets are annotated using a well-defined schema. Nevertheless, annotating persuasion techniques can be subjective. Therefore, models trained on these datasets might exhibit certain biases. Any biases found in the datasets are unintentional, and do not intend to harm any group or individual.

Acknowledgments. This work was partially supported by the European Union-NextGenerationEU, through the National Recovery and Resilience Plan of the Republic of Bulgaria [Grant Project No. BG-RRP-2.004-0008]. Views and opinions expressed are, however, those of the authors only and do not necessarily reflect those of the European Union. Neither the European Union nor the granting authority can be held responsible for them.

Declaration on Generative AI. During the preparation of this work, the authors used Grammarly to perform a grammar and spelling check. After using this tool/ service, the authors reviewed and edited the content as needed and take full responsibility for the publication's content.

References

1. Alam, F., Mubarak, H., Zaghouani, W., Da San Martino, G., Nakov, P.: Overview of the WANLP 2022 shared task on propaganda detection in Arabic. In: Bouamor, H., et al. (eds.) Proceedings of the Seventh Arabic Natural Language Processing Workshop, WANLP, pp. 108–118. ACL, Abu Dhabi, United Arab Emirates (Hybrid) (2022). https://doi.org/10.18653/v1/2022.wanlp-1.11, https://aclanthology.org/2022.wanlp-1.11/
2. Baleato Rodríguez, D., Dankers, V., Nakov, P., Shutova, E.: Paper bullets: modeling propaganda with the help of metaphor. In: Vlachos, A., Augenstein, I. (eds.) Findings of the Association for Computational Linguistics, EACL 2023, pp. 472–489. ACL, Dubrovnik, Croatia (2023). https://doi.org/10.18653/v1/2023.findings-eacl.35, https://aclanthology.org/2023.findings-eacl.35/
3. Barbaresi, A.: Trafilatura: a web scraping library and command-line tool for text discovery and extraction. In: Ji, H., Park, J.C., Xia, R. (eds.) Proceedings of the 59th Annual Meeting of the Association for Computational Linguistics and the 11th International Joint Conference on Natural Language Processing: System Demonstrations, pp. 122–131. ACL (2021). https://doi.org/10.18653/v1/2021.acl-demo.15, https://aclanthology.org/2021.acl-demo.15/
4. Barrón-Cedeño, A., Jaradat, I., Da San Martino, G., Nakov, P.: Proppy: organizing the news based on their propagandistic content. Inf. Process. Manage. **56**(5), 1849–1864 (2019). https://doi.org/10.1016/j.ipm.2019.03.005
5. Da San Martino, G., Barrón-Cedeño, A., Wachsmuth, H., Petrov, R., Nakov, P.: SemEval-2020 task 11: detection of propaganda techniques in news articles. In: Herbelot, A., Zhu, X., Palmer, A., Schneider, N., May, J., Shutova, E. (eds.) Proceedings of the Fourteenth Workshop on Semantic Evaluation, pp. 1377–1414. International Committee for Computational Linguistics, Barcelona (2020). https://doi.org/10.18653/v1/2020.semeval-1.186, https://aclanthology.org/2020.semeval-1.186/
6. Da San Martino, G., Shaar, S., Zhang, Y., Yu, S., Barrón-Cedeño, A., Nakov, P.: Prta: A system to support the analysis of propaganda techniques in the news. In: Celikyilmaz, A., Wen, T.H. (eds.) Proceedings of the 58th Annual Meeting of the Association for Computational Linguistics: System Demonstrations, pp. 287–293. ACL (2020). https://doi.org/10.18653/v1/2020.acl-demos.32, https://aclanthology.org/2020.acl-demos.32/

7. Da San Martino, G., Yu, S., Barrón-Cedeño, A., Petrov, R., Nakov, P.: Fine-grained analysis of propaganda in news articles. In: Inui, K., Jiang, J., Ng, V., Wan, X. (eds.) Proceedings of the 2019 Conference on Empirical Methods in Natural Language Processing and the 9th International Joint Conference on Natural Language Processing. pp. 5636–5646. EMNLP-IJCNLP, ACL, Hong Kong, China (2019). https://doi.org/10.18653/v1/D19-1565, https://aclanthology.org/D19-1565/
8. Devlin, J., Chang, M.W., Lee, K., Toutanova, K.: BERT: pre-training of deep bidirectional transformers for language understanding. In: Burstein, J., Doran, C., Solorio, T. (eds.) Proceedings of the 2019 Conference of the North American Chapter of the Association for Computational Linguistics: Human Language Technologies, Volume 1 (Long and Short Papers), pp. 4171–4186. ACL, Minneapolis, Minnesota (2019). https://doi.org/10.18653/v1/N19-1423, https://aclanthology.org/N19-1423/
9. Dimitrov, D., et al.: SemEval-2024 task 4: multilingual detection of persuasion techniques in memes. In: Ojha, A.K., Doğruöz, A.S., Tayyar Madabushi, H., Da San Martino, G., Rosenthal, S., Rosá, A. (eds.) Proceedings of the 18th International Workshop on Semantic Evaluation, SemEval-2024, pp. 2009–2026. ACL, Mexico City, Mexico (2024). https://doi.org/10.18653/v1/2024.semeval-1.275, https://aclanthology.org/2024.semeval-1.275/
10. Dimitrov, D., et al.: SemEval-2021 task 6: detection of persuasion techniques in texts and images. In: Palmer, A., Schneider, N., Schluter, N., Emerson, G., Herbelot, A., Zhu, X. (eds.) Proceedings of the 15th International Workshop on Semantic Evaluation, SemEval-2021, pp. 70–98. ACL (2021). https://doi.org/10.18653/v1/2021.semeval-1.7, https://aclanthology.org/2021.semeval-1.7/
11. Habernal, I., Hannemann, R., Pollak, C., Klamm, C., Pauli, P., Gurevych, I.: Argotario: computational argumentation meets serious games. In: Specia, L., Post, M., Paul, M. (eds.) Proceedings of the 2017 Conference on Empirical Methods in Natural Language Processing: System Demonstrations, pp. 7–12. ACL, Copenhagen, Denmark (2017). https://doi.org/10.18653/v1/D17-2002, https://aclanthology.org/D17-2002/
12. Hasanain, M., et al.: ArAIEval shared task: persuasion techniques and disinformation detection in Arabic text. In: Sawaf, H. (eds.) Proceedings of ArabicNLP 2023, pp. 483–493. ACL, Singapore (Hybrid) (2023). https://doi.org/10.18653/v1/2023.arabicnlp-1.44, https://aclanthology.org/2023.arabicnlp-1.44/
13. Huang, K.H., McKeown, K., Nakov, P., Choi, Y., Ji, H.: Faking fake news for real fake news detection: propaganda-loaded training data generation. In: Rogers, A., Boyd-Graber, J., Okazaki, N. (eds.) Proceedings of the 61st Annual Meeting of the Association for Computational Linguistics (Volume 1: Long Papers), pp. 14571–14589. ACL, Toronto, Canada (2023). https://doi.org/10.18653/v1/2023.acl-long.815, https://aclanthology.org/2023.acl-long.815/
14. IPA: Propaganda Analysis: Volume I of the Publications of the Institute for Propaganda Analysis, Inc, vol. 1. IPA, New York (1938)
15. Klie, J.C., Bugert, M., Boullosa, B., Eckart de Castilho, R., Gurevych, I.: The INCEpTION platform: machine-assisted and knowledge-oriented interactive annotation. In: Zhao, D. (ed.) Proceedings of the 27th International Conference on Computational Linguistics: System Demonstrations, pp. 5–9. ACL, Santa Fe, New Mexico (2018). https://aclanthology.org/C18-2002/

16. Nakov, P., Alam, F., Shaar, S., Da San Martino, G., Zhang, Y.: COVID-19 in Bulgarian social media: factuality, harmfulness, propaganda, and framing. In: Mitkov, R., Angelova, G. (eds.) Proceedings of the International Conference on Recent Advances in Natural Language Processing. pp. 997–1009. RANLP 2021, INCOMA Ltd., Held Online (2021). https://aclanthology.org/2021.ranlp-1.113/
17. Piskorski, J., et al.: Overview of the clef-2024 checkthat! lab task 3 on persuasion techniques. In: CEUR Workshop Proceedings, vol. 3740, Paper 26, pp. 299–310 (2024). https://ceur-ws.org/Vol-3740/paper-26.pdf
18. Piskorski, J., et al.: SlavicNLP 2025 shared task: detection and classification of persuasion techniques in parliamentary debates and social media. In: Piskorski, J., Přibáň, P., Nakov, P., Yangarber, R., Marcinczuk, M. (eds.) Proceedings of the 10th Workshop on Slavic Natural Language Processing, Slavic NLP 2025, pp. 254–275. ACL, Vienna, Austria (2025). https://doi.org/10.18653/v1/2025.bsnlp-1.27, https://aclanthology.org/2025.bsnlp-1.27/
19. Piskorski, J., Stefanovitch, N., Da San Martino, G., Nakov, P.: SemEval-2023 task 3: detecting the category, the framing, and the persuasion techniques in online news in a multi-lingual setup. In: Ojha, A.K., et al. (eds.) Proceedings of the 17th International Workshop on Semantic Evaluation, SemEval-2023, pp. 2343–2361. ACL, Toronto, Canada (2023). https://doi.org/10.18653/v1/2023.semeval-1.317, https://aclanthology.org/2023.semeval-1.317/
20. Rashkin, H., Choi, E., Jang, J.Y., Volkova, S., Choi, Y.: Truth of varying shades: analyzing language in fake news and political fact-checking. In: Palmer, M., Hwa, R., Riedel, S. (eds.) Proceedings of the 2017 Conference on Empirical Methods in Natural Language Processing, pp. 2931–2937. ACL, Copenhagen, Denmark (2017). https://doi.org/10.18653/v1/D17-1317, https://aclanthology.org/D17-1317/
21. Rivière, M., et al.: Gemma 2: improving open language models at a practical size. CoRR arXiv:2408.00118 (2024), https://doi.org/10.48550/arXiv.2408.00118
22. Snow, N., Jowett, G., O'Donnell, V.: Propaganda & Persuasion. SAGE Publications Inc, Thousand Oaks, USA, eight edn (2024)

A Survey on Multilingual Clinical Entity Linking in the LLM Age

Sylvia Vassileva(✉), Ivan Koychev, and Svetla Boytcheva

Faculty of Mathematics and Informatics, Sofia University "St. Kliment Ohridski", 5 J. Bourchier Blvd., Sofia, Bulgaria
{svasileva,koychev}@fmi.uni-sofia.bg, svetla@uni-sofia.bg

Abstract. Objective: Clinical entity linking is the task of associating spans in clinical text with unique concepts from standardized clinical ontologies. It is especially important in the clinical domain, given the complexity of terminology and the critical need for accurate interpretation by machine learning algorithms. Our goal is to survey methods for clinical entity linking across languages over the past three years, identify current trends in methodology, and analyze the impact of Large Language Models (LLMs) on solving this task.

Methods: We queried multiple paper databases for papers on clinical entity linking and normalization published between 2023 and 2025. We extracted and categorized the system and model architectures used in the papers, as well as the datasets, languages, entity types, and ontologies used for method validation.

Results: We analyzed 43 studies and found that the majority of methods use a retrieve-and-rerank architecture and employ transformer encoder models, with an increasing number of papers incorporating decoder models into the pipeline. We identified benchmark datasets for the task and the current state-of-the-art methods for each dataset. We discuss the multilingual strategies used by authors, including the use of multilingual models and neural machine translation from low-resource languages to high-resource languages such as English.

Conclusion: LLMs have been successfully applied to a wide range of tasks; however, applying them directly to entity linking has encountered challenges due to the large search space of clinical concepts. Incorporating LLMs into the entity linking pipeline has proven beneficial for candidate reranking and clinical term standardization. The lack of labeled datasets in different languages, as well as the complexity of clinical terms, remains a challenge for clinical entity linking.

Keywords: clinical entity linking · clinical entity normalization · LLM

1 Introduction

Entity linking, also known as normalization, is the task of associating spans in text with the unique concepts they represent from a standardized ontology. Once

I. K. Koychev (Ed.): ERIS 2025, PINNS 4, pp. 31–56, 2026.
https://doi.org/10.1007/978-3-032-30153-6_3

the spans (sometimes referred to as mentions) are linked to the ontology concept, the information can be used to automatically analyze multiple documents, filter and aggregate data, and identify similar documents based on the concepts within them. Entity linking is especially important in the biomedical and clinical domains, where many complex concepts exist, and matching them can help automate the processing of biomedical papers, clinical notes, and discharge summaries to support tasks such as drug discovery, biomedical question-answering, patient trial matching, and clinical decision support. Research on the clinical entity linking task has increased gradually since 2000, as shown in Fig. 1.

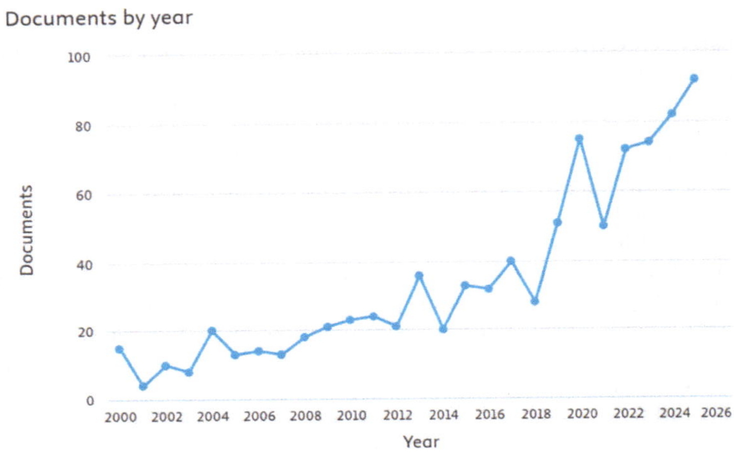

Fig. 1. Number of papers per year related to clinical entity linking or normalization indexed in Scopus from 2000 to 2025.

In this paper, we survey recent methods used for multilingual clinical entity linking in the last three years (2023–2025). We provide an overview of the tasks and key clinical ontologies and classification systems, as well as a comprehensive list of labeled clinical datasets suitable for training or as benchmarks for method comparison. We also review the different approaches and identify trends in the current clinical entity linking landscape. We compare results across benchmarks, identify the current state-of-the-art (SoTA) methods, and discuss the field's challenges and potential avenues for further research.

Existing surveys and the majority of datasets focus on biomedical texts and entity linking [14,30,83]. However, we focus our survey on clinical entity linking, since clinical and biomedical texts differ and therefore pose distinct challenges. Biomedical texts from papers and textbooks are usually well-structured, while clinical texts may not follow proper grammar, lack separate sentences, contain many typographical errors and abbreviations, and use different terminology depending on the clinician's specialty.

The primary goal of this survey is to review and analyze current trends and challenges in clinical entity linking in different languages. Since the release of ChatGPT[1], extensive research has been conducted to adapt Large Language Models (LLMs) to various clinical tasks; we also explore the impact of LLMs on clinical entity linking. We aim to study the following research questions (RQ):

- RQ1: Which methods and architectures are implemented most frequently for clinical entity linking?
- RQ2: What strategies can be used for adapting the methods for other languages, including low-resource settings?
- RQ3: To what extent can LLMs be successfully applied to clinical entity linking, and what are their strengths and limitations?

We queried multiple paper databases - Web of Science[2], Scopus[3], PubMed[4], Google Scholar[5], ArXiv[6], medRXiv[7], and HuggingFace[8] with the keywords "clinical entity linking" and "clinical entity normalization" and selected papers published in the period 2023–2025. We identified the main benchmark datasets used in these papers and searched for additional papers that used them in their experiments. For some datasets, we included older methods for comparison. In the case of classification for the International Classification of Diseases 10th Edition (ICD-10)[9], we include methods that treat it as an entity-linking problem and exclude methods that treat it as medical coding or concept indexing.

The contributions of this survey are as follows:

- A review of the latest clinical entity linking approaches in multiple languages, analysis of trends, and the most common approaches and adaptation strategies.
- An analysis of the impact of LLMs on clinical entity linking, their strengths and challenges in that context.

1.1 Task Description

Entity linking or normalization is the task of assigning a unique identity from a knowledge base to entities mentioned in text, thereby defining their meaning unambiguously in context. It is a crucial task in the automatic processing of information extracted from text. Entity linking usually relies on extracted mentions from the text by a previous step in the pipeline - entity detection.

[1] https://chat.openai.com/.
[2] https://www.webofscience.com/.
[3] http://scopus.com/.
[4] http://pubmed.ncbi.nlm.nih.gov/.
[5] https://scholar.google.com/.
[6] http://arxiv.org/.
[7] https://www.medrxiv.org/.
[8] https://huggingface.co/papers.
[9] https://icd.who.int/browse10/2019/en.

In clinical Natural Language Processing (NLP), entity linking is an essential component of processing clinical texts, grounding information in curated ontologies and systems such as the International Classification of Diseases 10th Edition (ICD-10)[10], Systematized Nomenclature of Medicine Clinical Terms (SNOMED-CT)[11], and Unified Medical Language System (UMLS)[12]. Figure 2 shows an example of entity linking to SNOMED-CT.

Another important task in the clinical domain is medical coding. It is the process of assigning ICD-10 statistical classification codes to a clinical document, such as a note or discharge summary [71]. Medical coding provides the basis for insurance reimbursement and statistical analysis and requires clinical expertise. It is an administrative task that requires effort and time, so automating it would help reduce the administrative burden on physicians. Efforts have been made to automate medical coding or to develop systems that facilitate human coders' work. Some medical coding methods use entity linking as part of their pipelines and will be examined in this study.

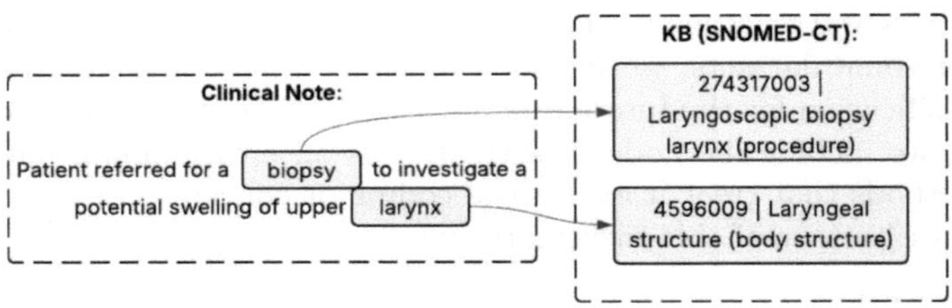

Fig. 2. The figure shows an example of entity linking in a clinical note.

1.2 Definitions

Knowledge Base: A knowledge base (KB) is a set of sentences in a knowledge representation language that represent assertions about the world, together with an inference mechanism that can derive new sentences from the existing ones [80]. Entities are the individual elements in a KB that are used to describe a domain. The KB consists of facts about entities expressed as predicates and relations.

Entity Linking: Given a set of mentions $M = \{m_1, .., m_n\}$ extracted from a document and a knowledge base KB with entities $E = \{e_1, .., e_k\}$, entity linking tries to find the mapping $f : M \mapsto E \cup \{NIL\}$ which assigns each mention to a single KB entity or the special NIL entity if no suitable entity exists in the KB [82]. A mention is a meaningful sequence of tokens from a document that represents a concept [82].

[10] https://icd.who.int/browse10/2019/en.
[11] https://www.snomed.org/.
[12] https://www.nlm.nih.gov/research/umls/index.html.

2 Data

Publicly available datasets for clinical entity normalization are scarce due to the sensitivity of clinical texts and the effort required to label them. However, some datasets are available for training, usually part of shared tasks.

2.1 Ontologies and Classification Systems

The datasets are labeled using standard medical ontologies and classification systems. The most commonly used systems in publicly available datasets are UMLS, SNOMED-CT, and ICD-10.

UMLS. Unified Medical Language System (UMLS)[13] [8] is an international system consisting of classifications, dictionaries, and coding standards for medical terms. The goal of UMLS is to facilitate the development and integration of biomedical information systems and services by standardizing and linking terms from different ontologies. UMLS consists of a meta-thesaurus that defines the mapping between different medical coding systems - about 5 million terms and their synonyms, a semantic network that defines semantic types and relationships, and specialized lexicons and tools. Semantic types are categories such as "disease or syndrome" or "drug". Semantic relations define important relationships between types, such as "drug" - "treats" - "disease or syndrome". UMLS has partial translations into various languages such as Spanish, German, French, Italian, and others[14].

SNOMED-CT. SNOMED-CT[15] is an ontology of medical terms for all specialties used internationally in electronic health records (EHRs) to standardize the diagnoses, symptoms, body structures, procedures, and medications. It is available in over 20 languages and can be mapped to other standard systems, such as ICD-10. SNOMED-CT consists of concepts, relationships, and descriptions, and provides synonyms for each concept in the available languages. It contains more than 360,000 concepts and more than 1.2 million terms and their synonyms. The ontology is updated regularly to reflect new terms relevant to clinical practice. SNOMED-CT provides definitions of concepts using their relationships with other concepts, and therefore, fully-defined concepts can be expressed using formal logic. For example, the concept Fracture of tibia is equivalent to "Fracture of lower limb" and "Injury of tibia" with "Associated morphology": "Fracture" and "Finding site": "Bone structure of tibia".

[13] https://www.nlm.nih.gov/research/umls/index.html.

[14] https://www.clinfowiki.org/wiki/index.php/Unified_Medical_Language_System.

[15] https://www.snomed.org/what-is-snomed-ct.

ICD-10. International Classification of Diseases (10th Revision) (ICD-10)[16] is an international standardized classification of diseases used in many countries for statistical purposes. ICD-10 is a hierarchical classification that organizes diseases into groups at several levels. Each disease is assigned a code for the group to which it belongs. Each ICD-10 code can range from 3 to 7 characters, depending on the group's specificity.

2.2 Datasets

Recent efforts have increased the number of public clinical datasets labeled for entity linking. English-language datasets predominate, but some data are also available in Spanish, German, French, and Greek. The most common ontologies/systems for labeling are SNOMED-CT, UMLS, and ICD-10. Datasets cover various entity types, including diseases, disorders, diagnoses, symptoms, procedures, and drugs. Table 1 lists the available datasets, including their features, and Table 4 shows statistics on the number of documents, mentions, and unique labeled concepts.

3 Evaluation Metrics

The entity linking task is usually evaluated by measuring how accurately mentions are mapped to the correct KB concepts, using classification metrics (F1, intersection over union) or ranking metrics (accuracy@k, recall@k) [51]. Classification metrics treat the pairs of mention-concept as the prediction target:

- Accuracy/accuracy@1/recall@1: Fraction of the mentions whose top predicted concept is the gold concept.
- Precision, recall, F1: Calculate the precision, recall, and F1 from overlap with the gold labels, where true positives are the correctly predicted pairs, false positives are incorrectly predicted pairs, and false negatives are pairs that were missed in predictions. Papers report micro- or macro-average F1 scores, where micro-average aggregates metrics across all mentions, while macro-average aggregates metrics per document, entity type, or concept.

Ranking metrics evaluate a ranked list of candidate concepts for each mention:

- Recall@k: The fraction of gold concepts that appear in the top k candidates.
- Mean Reciprocal Rank (MRR): For each mention, calculate the mean of the reciprocal rank of the first correct entity.
- Precision@k: Proportion of candidates in the top k that are correct. Precision@1 is identical to accuracy when exactly one prediction is returned.

In some cases, multiple concepts may be assigned to a mention, and all could be considered correct. In this case, the evaluation can be performed in a strict or relaxed manner. The strict approach counts the prediction as correct only if all predicted concepts match the ground-truth, while the relaxed approach requires only a single correct concept to be predicted to count it as correct [51].

[16] https://icd.who.int/browse10/2019/en#/.

Table 1. Clinical Entity Linking datasets statistics.

Dataset	Document type	Year	Main language	Entity Types	Ontology
DisTEMIST [69]	Clinical case reports	2022	Spanish	Disease	SNOMED-CT
MedProcNER [59]	Clinical case reports	2023	Spanish	Procedures	SNOMED-CT
SympTEMIST [58]	Clinical case reports	2023	Spanish	Symptoms	SNOMED-CT
SNOMED CT Entity Linking Challenge [25]	Discharge notes	2024	English	Findings, Procedures, Body structures	SNOMED-CT
MDACE [20]	Clinical records	2023	English	Diagnosis, Procedures	ICD-9, ICD-10
CUILESS [73]	Discharge summaries	2016	English	Disorder	SNOMED-CT
CodiEsp [70]	Clinical case reports	2020	Spanish	Diagnosis, Procedures	ICD-10
CanTEMIST [68]	Oncological case reports	2020	Spanish	Oncology	ICD-O
Bronco [52]	Discharge summaries	2021	German	Diagnosis, treatment, medication	ICD-10, OPS, ATC
MADE [46]	EHR notes	2019	English	ADE, Drug, Indication	MedDRA
ShARe/CLEF task 1 [75]	Discharge reports	2013	English	Disorder	UMLS
ElCardioCC [26]	Clinical case reports	2025	Greek	Diagnosis	ICD-10
n2c2 2019 [66]	Discharge summaries	2019	English	Problem, Test, Treatment	UMLS
ShARe corpus [76]	Discharge summaries	2015	English	Disorder	UMLS, SNOMED-CT
CLEF 2017 task 1 [39]	Cause of death notes	2017	French	Cause of death	ICD-10

4 Recent Methods

Table 2 shows a summary of recently developed systems for clinical entity linking and medical coding, and the different system and model architectures, as well as the base models. The majority of the systems use the approach of retrieve and rerank, consisting of a candidate generation step which reduces the search space and a subsequent reranking step to identify the most relevant concept based on the context [9,10,28,33,54,91,103]. Figure 3 shows the standard architecture with candidate generation and reranking steps.

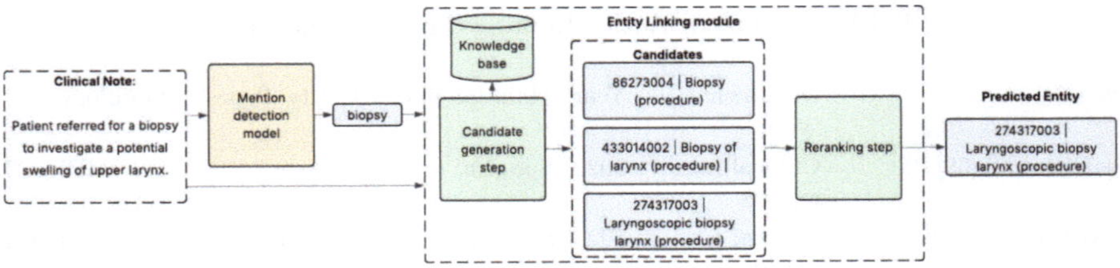

Fig. 3. The standard architecture used in entity linking systems consists of a retrieval step to generate candidates and a subsequent reranking step to produce the final prediction.

4.1 Encoder Models

Transformer encoders dominate model architectures, typically by training a bi-encoder to generate candidates and a cross-encoder to rerank them for the final prediction. The bi-encoder models are trained using a contrastive objective [60] so that the mention and concept embeddings are represented in the same vector space and mentions relating to a concept are close together, while the cross-encoder is trained for the classification task to predict which concept candidate is the best fit for the mention based on the context.

A common backbone model for this approach is SapBERT/SapBERT-XLMR [60,61]. In addition to the vector embeddings, some authors utilize term frequency-inverse document frequency (TF-IDF) vector representation to improve the candidate generation results [9,10]. Others use a sparse index for candidate generation, such as BM25 [35,57] rather than a transformer bi-encoder. Hou et al. [42] generate a multi-granularity mention-concept representation using character, word/phrase, and concept/dictionary inputs, trained with hard-negative sampling and a multi-similarity loss for entity linking in Chinese. Kreuzhaler et al. [53] use XLMR-SapBERT embeddings and k-nearest-neighbors (kNN) search to select the best ICD-10 code for disease descriptions provided by clinicians in German. Some authors use a simplified approach consisting only in a retrieval step and selection of top-1 candidate using a bi-encoder model pretrained on contrastive objective for the specific ontology [19,21,40,49,84,90,93,100].

A cross-encoder classifier is an important part of the system, as it allows selection of the best concept based on the mention's context. Multilingual models like SapBERT-XLMR [9,10,33] or language-specific monolingual models [2,28,32,42,101,103] are used to train the cross-encoder. Li et al. [57] use a stacking-BERT architecture that combines four BERT-based models and trains them as an ensemble for reranking using adversarial methods. Zhang et al. [101] fuse the information from a BERT-based mention encoder and a graph embedding representation of a Knowledge Graph to classify the syndrome in traditional Chinese medicine texts.

4.2 Decoder Models

Recently, more experiments are being performed using decoder models for entity linking, generating the mention's normalized name using in-context learning and supervised finetuning [3], or generating the mention's ICD-10 code using LLMs like GPT-4.1[17] and in-context learning [92]. Motzfeldt et al. [71] use a multi-agent system with LLMs to predict ICD-10 codes, which identify codeable mentions, and then try to link them using candidate generation from the ICD-10 index and a reranking step performed by an LLM to ensure the ICD-10 guidelines match the mention. Finally, a reconciler agent resolves any conflicts among the final ICD-10 codes predicted for the entire document.

Another multi-agent system for UMLS entity linking is proposed by Yang et al. [99]. They create separate agents for each step in the standard pipeline for clinical information extraction: section splitting, entity extraction, entity enhancement via linking, date normalization, attribute extraction, and final aggregation and formatting into the expected result schema. The entity linking module uses SapBERT-based retrieval and LLMs for disambiguation with GPT-o3-mini[18]. Garcia-Baragan et al. [35] use a two-stage approach for entity linking to UMLS oncology entities in breast cancer clinical notes, including BM25-based candidate retrieval and prompting GPT-4[19] with Chain-of-Thought (CoT) for candidate disambiguation. Finally, the prediction is checked against symbolic constraints from UMLS.

Large language models have also been used for candidate reranking and entity disambiguation. For example, Vassileva et al. [91] use GPT-3.5[20] to rerank candidates generated by SapBERT cosine similarity search, showing that the LLM's reranking step improves overall performance. Also, Gallego et al. [31] use GPT-4o[21] to rerank candidates generated by a trained bi-encoder, ClinLinker-KB, and provide reasoning for concept selection, which can help with system explainability.

Borchert et al. [11] tackle lexical variability in clinical text by using LLMs like GPT-4 to simplify and standardize the entity mention if SapBERT cosine similarity search is not able to retrieve candidates with a high similarity score. Their experiments show improved recall by 2.9pp during candidate generation, which also improves the overall system performance to 0.636 on the SympTEMIST dataset [58] using the gold-standard mentions.

Clinical narratives often contain many abbreviations and typographical errors and do not adhere to standard grammar. To address this issue, Abdulnazar et al. [1] use GPT-4 to clean and automatically rephrase the EHR document into a more readable and standardized form before proceeding with entity normalization of the terms. The linking is performed using bi-encoder-based candidate

[17] https://openai.com/index/gpt-4-1/.
[18] https://openai.com/index/openai-o3-mini/.
[19] https://openai.com/index/gpt-4/.
[20] https://openai.com/index/gpt-3-5-turbo-fine-tuning-and-api-updates/.
[21] https://openai.com/index/gpt-4o-system-card/.

generation with a large dictionary linked to SNOMED-CT, and the candidates are then reranked using GPT-4.

Puts et al. [77] use a novel approach for ICD-10 entity linking by expanding the textual mention using few-shot prompting, generating an ICD-10-like description of the clinical mention, and then using Retrieval-Augmented Generation (RAG) to retrieve ICD-10 candidates and rerank them using GPT-4. They find that the system underperforms in the retrieval component and suggest that using the ICD-10 alphabetic index could improve the retrieval recall and the overall system performance.

Wang et al. [95] use entity linking as a first step for multi-hop question answering in Chinese using LLMs (GLM4 [38]). They create a hybrid pipeline, starting with a dictionary-matching step using the Aho-Corasick algorithm [87], followed by optionally reducing the search space using predefined categories, candidate generation using RoBERTa [62] embeddings, and an LLM prompted to rerank and select the correct concept.

4.3 Other Methods

A lot of methods use a dictionary approach to link unambiguous entities as a first step, followed by a deep learning model to predict the concepts for the rest of the mentions [10,21,91–93]. However, Chen et al. [19] observe that dictionaries help in English but contribute very little in Chinese. InsightBuddy-AI, an application for medication extraction and linking to SNOMED-CT and ICD-10, uses fuzzy dictionary search for medication names to perform entity linking [79]. Schutz et al. [81] experiment with using selective Levenshtein similarity[22], abbreviation expansion, and alphabetic word ordering and find that it outperforms SapBERT encodings for ICD-10 entity linking in a small German dataset. Sousa et al. [85] use a UMLS dictionary lookup and filter by mention semantic type to perform entity linking of diseases, procedures, and drugs in Portuguese clinical diaries.

Classification is usually used in cases where the number of codes in the KB is limited - for example, Liu et al. [63] train a classifier encoder model to predict 324 ICD-10 codes for mentions in Greek clinical texts. This approach is very challenging when the full set of ICD-10 codes is the target of prediction, as the available training data does not sufficiently cover all codes. Lopez et al. [65] train a BERT-based classifier to predict the ICD-10 prefix and suffix separately, thus reducing the number of possible classes for each prediction. Tsujimura et al. [88] use a system trained via metric learning, using SciBERT [5] mention representations and a cosine-based output layer, with similarities to UMLS concepts as input to a softmax prediction layer. They note that overwriting the output layer's weights with the converged states of the hidden layers helps mitigate the classifier's bias toward the most common concepts. Their neural entity linking classifier achieves very good results on the MCN dataset [67], improving over the linear classifier's accuracy by 12pp.

[22] https://en.wikipedia.org/wiki/Levenshtein_distance.

4.4 Zero or Multiple Concepts

Handling the case where the mention concept does not exist in the KB can be modeled as predicting a NIL entity. This is commonly approached by adding a NIL entity candidate to the generated candidate list and training the reranking model to select the correct candidate. The NIL entity representation is usually a trainable embedding [28].

Sometimes multiple concepts or codes should be linked to a single mention, making it a multi-label problem. In this case, the entity linking task can be approached as a multi-label classification problem [43,64], selecting top candidates that pass a threshold [23], or as a multi-task learning problem to predict the number of labels associated with each mention [42].

A known challenge in clinical entity linking is the presence of composite mentions, which link to more than one concept, for example, "Degeneration of the left knee with loose body" combining "degeneration of the left knee" and "loose body in left knee" [29]. Fan et al. [29] introduce a mention decomposition step using ChatGPT, using in-context learning and providing standard terms and component types to guide the model to decompose the mention into the correct atomic mentions. Once the atomic mentions are identified, the authors link them using a bi-encoder recall model based on SBERT [78] and a cross-encoder candidate selector trained on atomic mentions, standard terms, and component types generated by ChatGPT. The system shows improvement in entity normalization on the Chinese CHIP-CDN dataset [102].

4.5 Methods Summary

RQ1: Which methods and architectures are implemented most frequently for clinical entity linking?

Figure 4 shows the distribution of system and model architectures used in recent papers. The majority of papers use a "Retrieve and rerank" approach - 43%, followed by simple retrieval - 32%. Classification (15%) and generation (9%) approaches are less preferred due to challenges posed by the large search space in clinical ontologies such as SNOMED-CT and UMLS. Transformer encoder models dominate the model architectures - about 60% of the papers use them as the main architecture in their systems. Decoders and dictionaries are the main approaches in about 9% and 13% of cases, respectively. Encoder and decoder combination is slightly more popular (15%), where an encoder is usually used to generate candidates and the decoder can rerank and disambiguate the mentions. Static vector encoders were used in just 2% of the papers as contextually-aware embeddings have shown improved performance in multiple downstream tasks.

4.6 Multilingual Strategies

RQ2: What Strategies can be Used for Adapting the Methods for Other Languages, Including Low-Resource Settings?

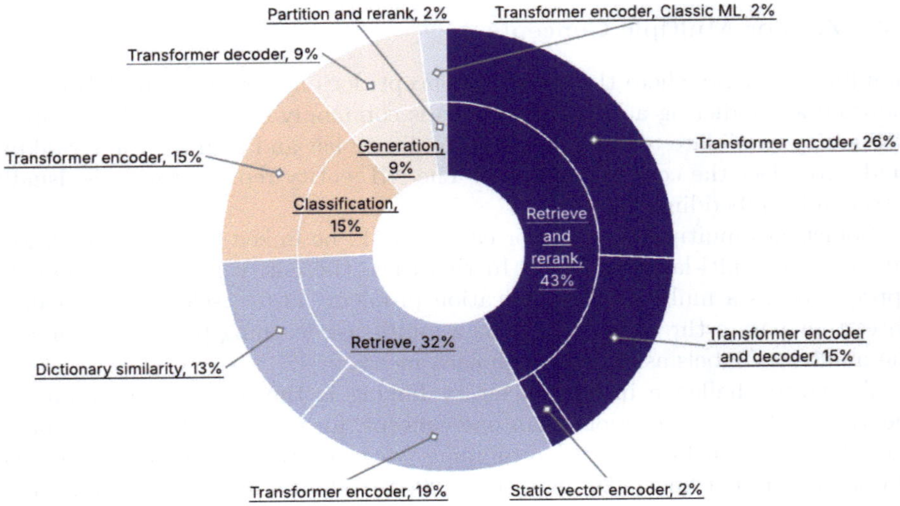

Fig. 4. Distribution of system and model architectures in papers.

In multilingual scenarios, there are often very few training examples available so authors use approaches based on pretrained multilingual models like SapBERT-XLMR [61] or neural machine translation to a more common language like English or Spanish using models like Helsinki-NLP/Opus-MT [86], GPT-3.5, Libre[23], Tigerbot-3 LLM [18], and EasyProject [9,17,19,49,63].

Pretraining models on raw text in an unsupervised manner has been shown to improve results on downstream tasks, including named entity recognition and entity linking [19]. D'Auria et al. [24] fine-tune a pre-trained bi-encoder and cross-encoder based on BLINK [97] for linking diseases and symptoms in Italian to DBPedia[24].

Borchert et al. [9] use an automatically translated dataset with label projection to train a cross-encoder, whose predictions are used as a regularization term during training of the final cross-encoder on gold-standard data. This approach combines a small gold-standard dataset with information derived from a machine-translated silver standard, thereby improving the system's performance across languages, with gains ranging from +12.6pp for Quaero (Medline) to +33.1pp for BRONCO (Treatments) [9].

Another method for tackling limited training data is to use KB information to augment concept representations and provide additional signal during linking. Gallego et al. [32] use the UMLS Knowledge Graph (KG) to enrich the candidate generation stage by providing parent relations for each candidate concept to the bi-encoder. This helps contrastive learning better differentiate concepts that are closer or farther apart in the KG hierarchy. In a subsequent study,

[23] https://www.libreoffice.org/.
[24] https://www.dbpedia.org/.

Table 2. Recent methods for clinical entity linking grouped by their architecture approaches. EL - Entity Linking, MC - Medical Coding.

Task	System Architecture	Model Architecture	Models	Papers
EL	Retrieve and rerank	Static vector encoder	FastText [50]	[6]
		Transformer encoder	SapBERT [60], SapBERT-XLMR [61], MedCPT [48], all-MiniLM-L12-v2[a], Spanish Biomedical RoBERTa[b], TeaBERT [16], SBERT [78], RoBERTa [62]	[9,10,12,28,33, 54] [2,24,29,42, 57,103]
		Transformer encoder and decoder	Mistral-7B-Instruct-v0.2 [47], GPT-3.5[c], GPT-4[d], GPT-4o[e]	MITEL-UNIUD[f], [1,11, 31,91,95,99]
EL	Retrieve	Dictionary similarity		[7] [22,37], KIRI[g], [81,85]
		Transformer encoder	SapBERT [60] SapBERT-XLMR [61], BGE-M3 [15], BlueBERT [74], PubMedBERT [41]	[19,21,40,49,53, 84,90,93,100]
MC	Partition and Rerank	Transformer encoder and ML Classifier	Spanish Biomedical RoBERTa	[4]
EL/MC	Generation	Transformer decoder	BioMistral [55], GPT-4, GPT-4.1[h], Qwen3-235B-A22B [98]	[3,71,77,92]
EL/MC	Classification	Transformer encoder	BETO [13], SciBERT [5], SapBERT [60], BETO-Galen[i], RoBERTa [62], mBERT-Galen[j]	[36,43,63–65,88,101]

[a] https://huggingface.co/sentence-transformers/all-MiniLM-L12-v2
[b] https://huggingface.co/PlanTL-GOB-ES/bsc-bio-es
[c] https://platform.openai.com/docs/models/gpt-3.5-turbo
[d] https://openai.com/index/gpt-4/
[e] https://openai.com/index/gpt-4o-system-card/
[f] https://drivendata.co/blog/snomed-ct-entity-linking-challenge-winners
[g] https://drivendata.co/blog/snomed-ct-entity-linking-challenge-winners
[h] https://openai.com/index/gpt-4-1/
[i] https://ieeexplore.ieee.org/document/9430499
[j] https://ieeexplore.ieee.org/document/9430499

the authors investigate enriching the reranking phase with KG information by fine-tuning BERT-based cross-encoders using enriched triplets in Spanish with parent-child information added to the negative examples [34]. Combining KB neighborhood information during candidate generation and reranking improves system performance, especially with codes absent from the training data.

5 Performance Comparison

We compare the performance of different models on the datasets reported in the respective papers and identify the highest reported scores. Table 3 summarizes the SoTA systems for each dataset and language across two settings: using the gold mentions directly or end-to-end, including mention detection and linking. Since the detection step introduces errors, the results reported in the end-to-end setting are lower; however, they better reflect the model's capabilities to tackle a real-world task. Different datasets measure different metrics, such as F1-score and IoU. Datasets which were released as part of a shared task, like the Spanish DisTEMIST [69], MedProcNER [59], CodiEsp [70], and SympTEMIST [58], English SNOMED CT Entity Linking Challenge [25] and Share/CLEF task1 [75], and Greek ElCardioCC [26], allow comparison of multiple different methods under the same settings which provides valuable information. End-to-end predictions are much harder because the entity detection step introduces errors in the mention boundary detection, or skips or adds mentions. State-of-the-art methods use dictionaries, retrieve and rerank with a bi-encoder and a cross-encoder, or an LLM for reranking, and train classifiers for a limited number of classes.

Table 3. SoTA performance on the different benchmark datasets in different languages. * is an average score across entity types.

Dataset	Language	Metric	SoTA w/gold mentions	SoTA end-to-end
SNOMED-CT Entity Linking Challenge [25]	en	IoU	–	0.4202 [25]
MDACE [20]	en	F1	0.6488 [3]	0.48 [43]
Share/Clef task1 2013 [75]	en	Acc	–	0.912 [28]
DisTEMIST [69]	es	F1	0.674 [33]	0.5657 [12]
MedProcNER [59]	es	F1	0.695 [33]	0.5707 [103]
CodiEsp-X [70]	es	F1	0.6764 [3]	0.611 [22,37]
SympTEMIST [58]	es	F1	0.6123 [91]	0.607 [10]
	en	F1	0.7525 [91]	–
	fr	F1	0.7284 [91]	–
	it	F1	0.726 [91]	–
	nl	F1	0.7365 [91]	–
	pt	F1	0.7271 [91]	–
ElCardioCC [26]	el	F1	0.856 [92]	0.6778 [63]
MedCT notes [19]	zh	IoU	–	0.30117 [19]
BRONCO150 [52]	de	F1	0.6816* [9]	–

6 Applications

Clinical entity linking is an important step in solving tasks related to clinical documentation structuring, clinical trial cohort discovery, and clinical decision support systems.

By mapping mentions to standardized clinical ontologies such as SNOMED-CT, ICD, and UMLS, EHR records can be searched, and data from different systems can be integrated for analysis. Entity linking can be used to automate ICD coding of notes and discharge summaries, thereby reducing physicians' administrative burden and improving billing accuracy [3,92].

Entity linking can also be used to generate patient representations comprising normalized diseases, drugs, and findings, which can be queried to support automatic cohort selection. The normalized terms can help align clinical trial eligibility criteria to patient data. LeafAI [27] uses a multi-stage pipeline to identify study-eligible patients within clinical databases by extracting and linking terms from EHRs to UMLS and constructing queries to retrieve eligible patients from the database. The authors demonstrate that their system outperforms an experienced human programmer at identifying patients eligible for clinical trials.

The linked concepts can be used for large-scale longitudinal analysis of patient data. Huang et al. [44] use entity linking as part of a comprehensive pipeline to generate reproducible, generalizable real-world evidence on disease outcomes from EHRs. The pipeline links EHR and registry data with labeled outcomes, then applies causal analysis to evaluate therapies for chronic diseases. LATTE also uses entity linking from clinical text to UMLS to create longitudinal visit embeddings that accurately annotate the timing of clinical events from longitudinal EHR data, leveraging limited labels efficiently [96]. Jafari et al. [45] examine methods and challenges in using EHRs for pharmacological research and provide examples of NLP methods that are used to extract and link medications to standardized terminologies for the purposes of constructing drug exposure cohorts, outcome ascertainment, and safety signal detection from clinical notes, and enabling multi-hospital studies by providing data interoperability.

7 Discussion

In this section, we discuss the challenges for clinical entity linking, analyze the usage of LLMs and their limitations for the task, and provide suggestions for future research. Challenges for entity linking stem from the complexity of clinical terminology, like the large search space for concepts, term ambiguity, code imbalance in datasets due to rarity of some diseases and conditions in the real world, and lack of expert translations of medical ontologies in low-resource languages. Further, due to patient privacy, it may not be possible to share datasets from some studies which makes the studies harder to reproduce and compare to others to determine their applicability in different hospital settings. More studies are using LLMs to generate synthetic datasets to allow method evaluation in different languages without risking patient data leaks.

7.1 Clinical Terminology Challenges

Clinical ontologies and classification systems are comprised of millions of concepts, making the search space large and difficult to navigate when identifying the correct concept. Studies often find a way to partition the full search space by selecting terms in certain categories - diseases, procedures, symptoms, or by classifying the terms in those categories during the named entity recognition (NER) step of the pipeline. Using retrieval algorithms further reduces the candidate set to a manageable number for ranking.

Another challenge for clinical entity linking is resolving term ambiguity in context to link a mention to the correct concept in the KB. Newman-Griffis et al. [72] examine the types of ambiguity observed across datasets to identify directions for future research. Their findings show that less than 15% of terms were ambiguous in standard datasets; however, over 50% were ambiguous in UMLS, highlighting the underrepresentation of this phenomenon in benchmarks. Developing datasets with improved representation of ambiguous terms would be helpful for future methods to address this problem [72].

Motzfeldt et al. [71] find that high-frequency ICD-10 codes are better handled with specialized discriminative models; however, the multi-agent LLM approach handles rare codes better and can easily support adding new codes without retraining. The authors argue that current systems are not ready to replace humans but should focus on assisting them in rare codes, guideline checking, and handling updates [71].

A fundamental challenge in cross-language clinical entity linking is the limited number of expert translations available for ontologies and systems such as SNOMED-CT and UMLS. Chen et al. [19] perform automatic translation for a subset ot SNOMED-CT and have experts validate the results; however, the effort required for a fully verified translation by experts is very expensive. Because ICD-10 is used internationally for insurance reimbursement in many countries, it is the most common classification system, with official translations in multiple languages; therefore, it is used in more studies.

7.2 Labeled Dataset Scarcity

Due to data privacy laws, clinical data cannot be shared without proper anonymization and appropriate reviews and approvals. This contributes to the scarcity of standard benchmark datasets in this area and makes it hard to compare different methods on the same data and under the same conditions. Some studies use in-house datasets to validate their models as they are developed within hospital projects. Since these models are usually not validated on standard benchmarks, it is challenging to determine whether they will perform well in different settings/hospitals. Future work can focus on reproducing and benchmarking published methods to identify the most effective approaches across different datasets.

The availability of training data in languages other than English is also a significant challenge for developing clinical methods and models. Recent shared-task efforts have provided public datasets for Spanish with labels for diseases, procedures, symptoms, and oncology, thereby facilitating advances in Spanish NLP methods. However, further work is needed for other languages that can benefit from machine translation and projection methods, while also requiring expert validation. With the development of LLMs, generating synthetic data is a promising approach to creating clinical datasets in different languages without exposing real patients' information.

7.3 LLM Strengths and Limitations

RQ3: To what Extent can LLMs be Successfully Applied to Clinical Entity Linking, and What are their Strengths and Limitations?

LLMs have been used successfully for reranking candidates generated by encoder models. LLMs can analyze mention context, including section- and document-level context, and select the best concept from a list of provided candidates. In addition, LLMs can be prompted to justify their selections or use CoT to provide explanations for physicians. Also, LLMs can be utilized to paraphrase mentions, enrich knowledge bases by generating concept descriptions, and generate examples for low-resource languages and rare codes. Current systems struggle with rare codes, so using LLMs to generate higher-quality examples of rare codes can help improve the performance of supervised systems [94].

LLMs struggle to generate concept identifiers end-to-end because the alphanumeric identifiers used in ontologies like UMLS and SNOMED-CT lack structure and semantics. Lee et al. [56] analyze the extent to which LLMs can understand and generate medical codes and find that they are prone to hallucinations and generate non-existent codes, making them unreliable for direct generation of concept codes. Newer LLM models, such as GPT-4.1, have shown promising results in generating ICD-10 codes [92]. ICD-10 is the most prevalent system used in datasets; its codes are short (3–4 characters) and hierarchical, which could help the model generate valid codes. However, the risk of hallucinations still remains a challenge, and additional post-processing or constrained generation should be performed to ensure valid codes are assigned.

Due to the large number of concepts in clinical systems and ontologies, it is hard to fit them in the LLM context and let the LLM select the correct one. Even if the token limit supports it, selecting one concept from hundreds of thousands is a needle-in-a-haystack problem that LLMs struggle with. As the transformer attention is distributed across more tokens, it becomes harder for the model to focus on the needle. Vandemoortele et al. [89] suggest that reducing the label space allows LLMs to focus their attention better on the relevant part of the context. Therefore, it is more beneficial to reduce the search space by retrieving relevant candidates using methods such as dense retrievers or dictionary-based similarity search, as most authors currently do. Another challenge with clinical ontologies is that they evolve continuously; therefore, relying only on memorized data during model training cannot predict new codes and concepts. This can be mitigated if the model has access to the latest version of the ontology and can efficiently query or retrieve relevant candidates using tools or other models.

In the future, constraint-based decoding strategies can be investigated to limit LLM predictions to the ontology or a subset of it, thereby mitigating hallucinations and large search-space issues. Recently, more multi-agent systems have been developed that incorporate entity linking as a step in processing clinical documents and use LLMs to rerank and validate predicted concepts. Future work in this area can help improve the clinical utility of entity linking systems.

8 Conclusion

Clinical entity linking is an important task in clinical NLP pipelines, as it is required to ground information in established medical concepts and facts. This paper provides an overview of the clinical entity linking task, categorizes recent methods for tackling it, and discusses challenges and directions for future work. LLMs have proven successful across a range of tasks; however, straightforward use for entity linking has encountered challenges due to the large search space of clinical concepts. Recently, an increasing number of studies use LLMs to simplify or standardize mention text, or to rerank candidates retrieved using embedding similarity within a RAG architecture. However, more research is needed to leverage LLM knowledge for accurate and reliable entity linking.

Acknowledgments. This work was partially supported by the European Union-NextGenerationEU, through the National Recovery and Resilience Plan of the Republic of Bulgaria [Grant Project No. BG-RRP-2.004-0008]. Views and opinions expressed are, however, those of the author only and do not necessarily reflect those of the European Union. Neither the European Union nor the granting authority can be held responsible for them.

Declaration on Generative AI. During the preparation of this work, the author(s) used Grammarly in order to: Grammar and spelling check. After using these tool(s)/ service(s), the author(s) reviewed and edited the content as needed and take(s) full responsibility for the publication's content.

A Dataset Statistics

Table 4. Clinical Entity Linking datasets statistics for available languages.

Dataset	Language	# documents	# mentions	# unique concepts
DisTEMIST [69]	Spanish	1000	10665	7303
	English	1000	7775	2749
	Portuguese	1000	7634	2528
	Catalan	1000	9965	3762
	Italian	1000	7526	2771
	French	1000	7553	2620
	Romanian	1000	5045	2090
MedProcNER [59]	Spanish	1000	14684	7179
SympTEMIST [58]	Spanish	1000	12196	1534
	Catalan	1000	3043	1307
	Czech	1000	722	395
	English	1000	2003	841
	Portuguese	1000	1686	671
	French	1000	1802	780
	Italian	1000	1924	823
	Dutch	1000	1570	645
	Romanian	1000	1438	656
	Swedish	1000	1399	609
SNOMED CT Entity Linking Challenge [25]	English	272	74808	6624
MDACE [20]	English	1192	9497	1758
CUILESS [73]	English	431	5397	1738
CodiEsp [70]	Spanish	1000	18435	3427
CanTEMIST [68]	Spanish	1301	16020	850
Bronco [52]	German	200	11124	4027
MADE [46]	English	1089	43000	–
ShARe/CLEF task 1 [75]	English	431	19557	1871
ElCardioCC [26]	Greek	1000	15864	324
n2c2 2019 [66]	English	100	10919	3792
ShARe corpus [76]	English	531	19111	2491
CLEF 2017 task 1 [39]	French	125384	125384	3723
	English	19995	19995	1413

References

1. Abdulnazar, A., Roller, R., Schulz, S., Kreuzthaler, M.: Large language models for clinical text cleansing enhance medical concept normalization. IEEE Access (2024)
2. Alhassan, A., Schlegel, V., Aloud, M., Batista-Navarro, R., Nenadic, G.: Generative models and sentence transformers for the recognition and normalization of continuous and discontinuous phenotype mentions: Model development and evaluation. JMIR Med. Inform. **13**, e68558 (2025)

3. Barreiros, L., Coutinho, I., Correia, G., Martins, B.: Explainable icd coding via entity linking. In: Proceedings of the Second Workshop on Patient-Oriented Language Processing (CL4Health), pp. 219–227 (2025)
4. Barros, J., Rojas, M., Dunstan, J., Abeliuk, A.: Divide and conquer: an extreme multi-label classification approach for coding diseases and procedures in spanish. In: Proceedings of the 13th International Workshop on Health Text Mining and Information Analysis (LOUHI), pp. 138–147 (2022)
5. Beltagy, I., Lo, K., Cohan, A.: Scibert: a pretrained language model for scientific text. arXiv preprint arXiv:1903.10676 (2019)
6. Bernik, M., Tovornik, R., Fabjan, B., et al.: Diagoza: a natural language processing tool for automatic annotation of clinical free text with snomed-ct (2022)
7. Blanco, A., Pérez, A., Casillas, A.: Ixa-aaa at clef ehealth 2020 codiesp. automatic classification of medical records with multi-label classifiers and similarity match coders. In: CLEF (Working Notes) (2020)
8. Bodenreider, O.: The unified medical language system (umls): integrating biomedical terminology. Nucleic Acids Res. **32**(Database-Issue), 267–270 (2004). http://dblp.uni-trier.de/db/journals/nar/nar32.html#Bodenreider04
9. Borchert, F., Llorca, I., Roller, R., Arnrich, B., Schapranow, M.P.: xmen: a modular toolkit for cross-lingual medical entity normalization. JAMIA Open **8**(1), ooae147 (2025)
10. Borchert, F., Llorca, I., Schapranow, M.P.: Hpi-dhc@ bc8 symptemist track: detection and normalization of symptom mentions with spanmarker and xmen. Islamaj, R., Arighi, C., Campbell, I., et al (2023)
11. Borchert, F., Llorca, I., Schapranow, M.P.: Improving biomedical entity linking for complex entity mentions with llm-based text simplification. Database **2024**, baae067 (2024)
12. Borchert, F., Schapranow, M.P.: Hpi-dhc@ bioasq distemist: Spanish biomedical entity linking with pre-trained transformers and cross-lingual candidate retrieval. In: CLEF (Working Notes), pp. 244–258 (2022)
13. Cañete, J., Chaperon, G., Fuentes, R., Ho, J.H., Kang, H., Pérez, J.: Spanish pre-trained bert model and evaluation data. arXiv preprint arXiv:2308.02976 (2023)
14. Chen, H., Zhou, Y., Li, R., Illa, A.M., Cleveland, A., Ding, J.: A comprehensive survey on medical concept normalization: datasets, techniques, applications, and future directions. Tech. Appl. Future Direct. (2025)
15. Chen, J., Xiao, S., Zhang, P., Luo, K., Lian, D., Liu, Z.: Bge m3-embedding: multi-lingual, multi-functionality, multi-granularity text embeddings through self-knowledge distillation. arXiv preprint arXiv:2402.03216 (2024)
16. Chen, L., Qi, Y., Wu, A., Deng, L., Jiang, T.: Teabert: an efficient knowledge infused cross-lingual language model for mapping chinese medical entities to the unified medical language system. IEEE J. Biomed. Health Inform. **27**(12), 6029–6038 (2023)
17. Chen, Y., Jiang, C., Ritter, A., Xu, W.: Frustratingly easy label projection for cross-lingual transfer. In: Findings of the Association for Computational Linguistics: ACL 2023, pp. 5775–5796 (2023)
18. Chen, Y., Cai, W., Wu, L., Li, X., Xin, Z., Fu, C.: Tigerbot: an open multilingual multitask llm. arXiv preprint arXiv:2312.08688 (2023)
19. Chen, Y., Huang, D., Shen, Y., Xu, H., Sheng, L., Zhou, Q., Wang, K.: Medct: a clinical terminology graph for generative ai applications in healthcare. In: International Conference on Computational Science, pp. 24–38. Springer (2025)
20. Cheng, H., et al.: Mdace: mimic documents annotated with code evidence (2023). arXiv:2307.03859

21. Chizhikova, M., et al.: Coming a long way with pre-trained transformers and string matching techniques: clinical procedure mention recognition and normalization. In: CLEF (Working Notes), pp. 91–101 (2023)
22. Cossin, S., Jouhet, V.: Iam at clef ehealth 2020: concept annotation in spanish electronic health records. In: CLEF (Working Notes) (2020)
23. Cuffy, C., French, E., Fehrmann, S., McInnes, B.T.: Exploring representations for singular and multi-concept relations for biomedical named entity normalization. In: Companion Proceedings of the Web Conference 2022, pp. 823–832 (2022)
24. D'Auria, D., Moscato, V., Postiglione, M., Romito, G., Sperlí, G.: Improving graph embeddings via entity linking: a case study on italian clinical notes. Intell. Syst. Appl. **17**, 200161 (2023)
25. Davidson, R., et al.: Snomed ct entity linking challenge. J. Am. Med. Inf. Assoc. **32**(9), 1397–1406 (2025). https://doi.org/10.1093/jamia/ocaf104
26. Dimitriadis, D., et al.: Overview of elcardiocc task on clinical coding in cardiology at bioasq 2025. In: CLEF (2025)
27. Dobbins, N.J., et al.: Leafai: query generator for clinical cohort discovery rivaling a human programmer. J. Am. Med. Inform. Assoc. **30**(12), 1954–1964 (2023). https://doi.org/10.1093/jamia/ocad149
28. Dong, H., Chen, J., He, Y., Liu, Y., Horrocks, I.: Reveal the unknown: out-of-knowledge-base mention discovery with entity linking. In: Proceedings of the 32nd ACM International Conference on Information and Knowledge Management, pp. 452–462 (2023)
29. Fan, Y., Zhu, Y., Xue, K., Liu, J., Ruan, T.: Rrnorm: a novel framework for chinese disease diagnoses normalization via llm-driven terminology component recognition and reconstruction. In: Findings of the Association for Computational Linguistics ACL 2024, pp. 9162–9175 (2024)
30. French, E., McInnes, B.T.: An overview of biomedical entity linking throughout the years. J. Biomed. Inform. **137**, 104252 (2023)
31. Gallego, F., et al.: Context-aware reasoning with large language models for medical entity linking in spanish. Authorea Preprints (2025)
32. Gallego, F., López-García, G., Gasco, L., Krallinger, M., Veredas, F.J.: Clinlinker-kb: clinical entity linking in spanish with knowledge-graph enhanced biencoders. Available at SSRN 4939986 (2024)
33. Gallego, F., López-García, G., Gasco-Sánchez, L., Krallinger, M., Veredas, F.J.: Clinlinker: medical entity linking of clinical concept mentions in spanish. In: International Conference on Computational Science, pp. 266–280. Springer (2024)
34. Gallego, F., Ruas, P., Couto, F.M., Veredas, F.J.: Enhancing cross-encoders using knowledge graph hierarchy for medical entity linking in zero-and few-shot scenarios. Knowl.-Based Syst. **314**, 113211 (2025)
35. García-Barragán, Á., et al.: Nssc: a neuro-symbolic ai system for enhancing accuracy of named entity recognition and linking from oncologic clinical notes. Med. Biol. Eng. Comput. **63**(3), 749–772 (2025)
36. García-Pablos, A., Perez, N., Cuadros, M.: Vicomtech at cantemist 2020. In: Proceedings of the Iberian Languages Evaluation Forum (IberLEF 2020), CEUR Workshop Proceedings, vol. 17, p. 25 (2020)
37. García-Santa, N., Cetina, K., Cappellato, L., Eickhoff, C., Ferro, N., Nevéol, A.: Fle at clef ehealth 2020: text mining and semantic knowledge for automated clinical encoding. In: CLEF (Working Notes) (2020)
38. GLM, T., et al.: Chatglm: a family of large language models from glm-130b to glm-4 all tools. arXiv preprint arXiv:2406.12793 (2024)

39. Goeuriot, L., et al.: Clef 2017 ehealth evaluation lab overview. In: Jones, G.J., et al. (eds.) Experimental IR Meets Multilinguality, Multimodality, and Interaction, pp. 291–303. Springer International Publishing, Cham (2017)
40. Grazhdanski, G., Vassileva, S., Koychev, I., Boytcheva, S.: Team fusion@ su@ bc8 symptemist track: transformer-based approach for symptom recognition and linking. New Orleans, USA (2023)
41. Gu, Y., et al.: Domain-specific language model pretraining for biomedical natural language processing. ACM Trans. Comput. Healthcare (HEALTH) **3**(1), 1–23 (2021)
42. Hou, W., He, K., Mao, R., Wang, J., Wang, X., Feng, M.: Multi-granularity semantic extraction and multi-task fusion for chinese medical entity normalization. Expert Syst. Appl. 129088 (2025)
43. Huang, C.W., Tsai, S.C., Chen, Y.N.: Plm-icd: automatic icd coding with pretrained language models. arXiv preprint arXiv:2207.05289 (2022)
44. Huang, F., et al.: Advancing the use of longitudinal electronic health records: Tutorial for uncovering real-world evidence in chronic disease outcomes. J. Med. Internet Res. **27**, e71873 (2025)
45. Jafari, E., et al.: Using electronic health records for clinical pharmacology research: challenges and considerations. Clin. Transl. Sci. **17**(7), e13871 (2024)
46. Jagannatha, A., Liu, F., Liu, W., Yu, H.: Overview of the first natural language processing challenge for extracting medication, indication, and adverse drug events from electronic health record notes (made 1.0). Drug Saf. **42**(1), 99–111 (2019)
47. Jiang, A.Q., et al.: Mistral 7b (2023). arXiv:2310.06825
48. Jin, Q., et al.: Medcpt: contrastive pre-trained transformers with large-scale pubmed search logs for zero-shot biomedical information retrieval. Bioinformatics **39**(11), btad651 (2023)
49. Jonker, R.A., Almeida, T., Matos, S., Almeida, J.: Team bit. ua@ bc8 symptemist track: a two-step pipeline for discovering and normalizing clinical symptoms in spanish. Islamaj, R., Arighi, C., Campbell, I. et al. (2023)
50. Joulin, A., Grave, E., Bojanowski, P., Douze, M., Jégou, H., Mikolov, T.: Fasttext. zip: compressing text classification models. arXiv preprint arXiv:1612.03651 (2016)
51. Kartchner, D., et al.: A comprehensive evaluation of biomedical entity linking models. In: Proceedings of the 2023 Conference on Empirical Methods in Natural Language Processing, pp. 14462–14478 (2023)
52. Kittner, M., et al.: Annotation and initial evaluation of a large annotated german oncological corpus. JAMIA Open **4**(2), ooab025 (2021). https://doi.org/10.1093/jamiaopen/ooab025
53. Kreuzthaler, M., Pfeifer, B., Schulz, S.: Secondary use of clinical problem list descriptions for bi-encoder based icd-10 classification. In: AMIA Annual Symposium Proceedings, vol. 2024, p. 620 (2025)
54. Kulyabin, M., Sokolov, G., Galaida, A., Maier, A., Arias-Vergara, T.: Snobert: a benchmark for clinical notes entity linking in the snomed ct clinical terminology. In: International Conference on Pattern Recognition, pp. 154–163. Springer (2024)
55. Labrak, Y., Bazoge, A., Morin, E., Gourraud, P.A., Rouvier, M., Dufour, R.: Biomistral: a collection of open-source pretrained large language models for medical domains. arXiv preprint arXiv:2402.10373 (2024)
56. Lee, S.A., Lindsey, T.: Do large language models understand medical codes? CoRR (2024)

57. Li, L., Zhai, Y., Gao, J., Wang, L., Hou, L., Zhao, J.: Stacking-bert model for chinese medical procedure entity normalization. Math. Biosci. Eng. **20**(1), 1018–1036 (2023)
58. Lima-López, S., Farré-Maduell, E., Gasco-Sánchez, L., Rodríguez-Miret, J., Krallinger, M.: Overview of symptemist at biocreative viii: corpus, guidelines and evaluation of systems for the detection and normalization of symptoms, signs and findings from text
59. Lima-López, S., et al.: Overview of MedProcNER task on medical procedure detection and entity linking at BioASQ 2023. In: Working Notes of CLEF 2023 – Conference and Labs of the Evaluation Forum (2023)
60. Liu, F., Shareghi, E., Meng, Z., Basaldella, M., Collier, N.: Self-alignment pretraining for biomedical entity representations. In: Toutanova, K., et al. (eds.) Proceedings of the 2021 Conference of the North American Chapter of the Association for Computational Linguistics: Human Language Technologies, pp. 4228–4238. Association for Computational Linguistics (2021). https://doi.org/10.18653/v1/2021.naacl-main.334, https://aclanthology.org/2021.naacl-main.334/
61. Liu, F., Vulić, I., Korhonen, A., Collier, N.: Learning domain-specialised representations for cross-lingual biomedical entity linking. In: Zong, C., Xia, F., Li, W., Navigli, R. (eds.) Proceedings of the 59th Annual Meeting of the Association for Computational Linguistics and the 11th International Joint Conference on Natural Language Processing (Volume 2: Short Papers), pp. 565–574. ACL (2021). https://doi.org/10.18653/v1/2021.acl-short.72, https://aclanthology.org/2021.acl-short.72/
62. Liu, Y., et al.: Roberta: a robustly optimized bert pretraining approach. arXiv preprint arXiv:1907.11692 (2019)
63. Liu, Y., Zhu, S.: Lyx_dmiip_fdu at bioasq 2025: Utilizing bert embeddings for biomedical text mining. In: CLEF (2025)
64. López-García, G., Jerez, J.M., Ribelles, N., Alba, E., Veredas, F.J.: Transformers for clinical coding in spanish. IEEE Access **9**, 72387–72397 (2021)
65. López-García, G., Jerez, J.M., Ribelles, N., Alba, E., Veredas, F.J.: Explainable clinical coding with in-domain adapted transformers. J. Biomed. Inform. **139**, 104323 (2023)
66. Luo, Y.F., Sun, W., Rumshisky, A.: Mcn: a comprehensive corpus for medical concept normalization. J. Biomed. Inform. **92**, 103132 (2019). https://doi.org/10.1016/j.jbi.2019.103132, https://www.sciencedirect.com/science/article/pii/S1532046419300504
67. Luo, Y.F., Sun, W., Rumshisky, A.: Mcn: a comprehensive corpus for medical concept normalization. J. Biomed. Inform. **92**, 103132 (2019)
68. Miranda-Escalada, A., Farré, E., Krallinger, M.: Named entity recognition, concept normalization and clinical coding: overview of the cantemist track for cancer text mining in spanish, corpus, guidelines, methods and results. In: Proceedings of the Iberian Languages Evaluation Forum (IberLEF 2020), CEUR Workshop Proceedings (2020)
69. Miranda-Escalada, A., et al.: Overview of distemist at bioasq: automatic detection and normalization of diseases from clinical texts: results, methods, evaluation and multilingual resources (2022)
70. Miranda-Escalada, A., Gonzalez-Agirre, A., Armengol-Estapé, J., Krallinger, M.: Overview of automatic clinical coding: annotations, guidelines, and solutions for non-english clinical cases at codiesp track of clef ehealth 2020. In: CLEF (Working Notes), vol. 2020 (2020)

71. Motzfeldt, A., Edin, J., Christensen, C.L., Hardmeier, C., Maaløe, L., Rogers, A.: Code like humans: a multi-agent solution for medical coding, pp. 22612–22627 (2025)
72. Newman-Griffis, D., Divita, G., Desmet, B., Zirikly, A., Rosé, C.P., Fosler-Lussier, E.: Ambiguity in medical concept normalization: an analysis of types and coverage in electronic health record datasets. J. Am. Med. Inform. Assoc. **28**(3), 516–532 (2021)
73. Osborne, J.D., Neu, M.B., Danila, M.I., Solorio, T., Bethard, S.J.: CUILESS2016: a clinical corpus applying compositional normalization of text mentions. J. Biomed. Semant. **9**(1), 2:1–2:9 (2018). https://doi.org/10.1186/S13326-017-0173-6
74. Peng, Y., Yan, S., Lu, Z.: Transfer learning in biomedical natural language processing: an evaluation of bert and elmo on ten benchmarking datasets. In: Proceedings of the 2019 Workshop on Biomedical Natural Language Processing (BioNLP 2019), pp. 58–65 (2019)
75. Pradhan, S., et al.: Task 1: Share/clef ehealth evaluation lab 2013. CLEF (working notes), vol. 1179 (2013)
76. Pradhan, S., et al.: Evaluating the state of the art in disorder recognition and normalization of the clinical narrative. J. Am. Med. Inf. Assoc. **22**(1), 143–154 (2014). https://doi.org/10.1136/amiajnl-2013-002544
77. Puts, S., Zegers, C.M., Dekker, A., Bermejo, I.: Developing an icd-10 coding assistant: pilot study using roberta and gpt-4 for term extraction and description-based code selection. JMIR Formative Res. **9**, e60095 (2025)
78. Reimers, N., Gurevych, I.: Sentence-bert: Sentence embeddings using siamese bert-networks. arXiv preprint arXiv:1908.10084 (2019)
79. Romero, P., Han, L., Nenadic, G.: Insightbuddy-ai: medication extraction and entity linking using large language models and ensemble learning. arXiv preprint arXiv:2409.19467 (2024)
80. Russell, S., Norvig, P.: Artificial Intelligence: A Modern Approach, 3rd edn. Prentice Hall, Upper Saddle River, New Jersey (2010)
81. Schulz, S., Abdulnazar, A., Kreuzthaler, M.: Clustering similar diagnosis terms, pp. 837–838 (2023)
82. Shen, W., Wang, J., Han, J.: Entity linking with a knowledge base: Issues, techniques, and solutions. IEEE Trans. Knowl. Data Eng. **27**(2), 443–460 (2015). https://doi.org/10.1109/TKDE.2014.2327028
83. Shi, J., Yuan, Z., Guo, W., Ma, C., Chen, J., Zhang, M.: Knowledge-graph-enabled biomedical entity linking: a survey. World Wide Web **26**(5), 2593–2622 (2023)
84. Singh, A., Krishnamoorthy, S., Ortega, J.E.: Neighbert: medical entity linking using relation-induced dense retrieval. J. Healthcare Inf. Res. **8**(2), 353–369 (2024)
85. Sousa, H., Mario Jorge, A., Pasquali, A., Santos, C., Lopes, M.: A biomedical entity extraction pipeline for oncology health records in portuguese. In: Proceedings of the 38th ACM/SIGAPP Symposium on Applied Computing, pp. 950–956 (2023)
86. Tiedemann, J., Thottingal, S.: Opus-mt–building open translation services for the world. In: Annual Conference of the European Association for Machine Translation, pp. 479–480. European Association for Machine Translation (2020)
87. Tseng, K.K., Chen, H., Chen, C., Bansong, C.: A secure live signature verification with aho-corasick histogram algorithm for mobile smart pad. Electronics **10**(11), 1337 (2021)
88. Tsujimura, T., Miwa, M., Sasaki, Y.: Large-scale neural biomedical entity linking with layer overwriting. J. Biomed. Inform. **143**, 104433 (2023)

89. Vandemoortele, N., Steenwinckel, B., Ongenae, F., Van Hoecke, S.: From haystack to needle: label space reduction for zero-shot classification. arXiv preprint arXiv:2502.08436 (2025)
90. Vassileva, S., Grazhdanski, G., Boytcheva, S., Koychev, I.: Fusion@ bioasq medprocner: Transformer-based approach for procedure recognition and linking in spanish clinical text. In: CLEF (Working Notes), pp. 190–205 (2023)
91. Vassileva, S., Grazhdanski, G., Koychev, I., Boytcheva, S.: Transformer-based approach for symptom recognition and multilingual linking. Database **2024**, baae090 (2024)
92. Vassileva, S., Koychev, I., Boytcheva, S.: Using llms for multilingual clinical entity linking to icd-10. arXiv preprint arXiv:2509.04868 (2025)
93. Velichkov, B., Datseris, A., Vassileva, S., Boytcheva, S.: Enigma@ elcardiocc: bridging ner and icd-10 entity linking-a hybrid method for greek clinical narratives. In: CLEF (2025)
94. Vo, T., Wu, W., Ding, K.: Synthetic clinical notes for rare icd codes: a data-centric framework for long-tail medical coding. arXiv preprint arXiv:2511.14112 (2025)
95. Wang, F., et al.: Llm-kgmqa: large language model-augmented multi-hop question-answering system based on knowledge graph in medical field. Knowl. Inf. Syst. 1–43 (2025)
96. Wen, J., et al.: Latte: label-efficient incident phenotyping from longitudinal electronic health records. Patterns **5**(1) (2024)
97. Wu, L., Petroni, F., Josifoski, M., Riedel, S., Zettlemoyer, L.: Scalable zero-shot entity linking with dense entity retrieval. In: Proceedings of the 2020 Conference on Empirical Methods in Natural Language Processing (EMNLP), pp. 6397–6407 (2020)
98. Yang, A., et al.: Qwen3 technical report. arXiv preprint arXiv:2505.09388 (2025)
99. Yang, Z., et al.: Clines: clinical llm-based information extraction and structuring agent. medRxiv, pp. 2025–12 (2025)
100. Zhai, W., Jiang, R., Huang, X., Bian, J., Zhu, S.: Investigation of synonym expansion and self-alignment pretraining for enhancing human phenotype ontology concept recognition. Artif. Intell. Med. 103282 (2025)
101. Zhang, G., Jiang, H., Kuai, L., Li, B., Huang, C., Fei, X., Huang, Z.: Kgsd-net: a knowledge graph syndrome differentiation network for syndrome classification. Front. Med. **12**, 1555781 (2025)
102. Zhang, N., et al.: Cblue: a chinese biomedical language understanding evaluation benchmark. In: Proceedings of the 60th Annual Meeting of the Association for Computational Linguistics (volume 1: long papers), pp. 7888–7915 (2022)
103. Zotova, E., Pablos, A.G., Cuadros, M., Rigau, G.: Vicomtech at medprocner 2023: Transformers-based sequence-labelling and cross-encoding for entity detection and normalisation in spanish clinical texts. In: CLEF (Working Notes), pp. 206–218 (2023)

EEG Data Analysis for Dementia State Recognition

Nadezhda Frantseva[1] and Olga Georgieva[1,2](✉)

[1] Faculty of Mathematics and Informatics, Sofia University "St. Kliment Ohridski", Sofia, Bulgaria
{franceva,o.georgieva}@fmi.uni-sofia.bg
[2] GATE Institute, Sofia University "St. Kliment Ohridski", Sofia, Bulgaria

Abstract. Degenerative diseases are the most common dementias, especially among the elderly population. Despite advances in neuroscience, there is still no answer to the cause of these diseases. A correct diagnosis made in the early stages of the disease is of great importance, since the treatment of the different types of dementia differs. In response to such a demand, the present article aims to propose a procedure to distinguish patients with Alzheimer's, frontotemporal dementia, and healthy people based on their resting state EEG data. The procedure relays rigorous feature selection and follows clustering analysis to define the three groups. The distinguishing ability is assessed through clustering quality metrics. The results show that the use of statistically significant features selected by the Wilcoxon test of channel coherence leads to very good group discrimination, according to clustering indexes. The division of the beta frequency band into three subbands - low, medium and high range, allows for a more precise accounting of neurophysiological differences between groups. Additional division of the alpha band also increases the ability to group compared to using only the five standard frequency bands. A significant improvement in the ability to find the three groups was achieved using a semi-supervised learning approach by identifying good initial values of the cluster centers that are close to their extreme values.

Keywords: EEG data analysis · Feature selection · Clustering analysis · Dementia

1 Introduction

Dementia is a disease with a progressive deterioration of brain activity and consequently behavioral and social skills, which affects a person's ability to live independently. Degenerative diseases are the most common dementias, especially among the elderly population. Despite advances in neuroscience, there is still no answer to the cause of these diseases. A correct diagnosis made in the early stages of the disease is of great importance, since the treatment of the different types of dementia differs.

I. K. Koychev (Ed.): ERIS 2025, PINNS 4, pp. 57–72, 2026.
https://doi.org/10.1007/978-3-032-30153-6_4

One very promising approach in this direction is the analysis of Electroencephalography signals (EEG). EEG measures brain activity in specific areas (brain channels). It is a non-invasive method for recording the electrical activity of the cerebral cortex, which provides valuable information about the state and functioning of the brain in real time. The other value of EEG data to underline is that they can provide indications of brain abnormalities that cannot be detected by other methods such as imaging or patient behavior [1,2].

The diagnosis of cognitive disorders by EEG is based on the detection of characteristic changes in the spectral, temporal, and functional structure of the signals [14]. In Alzheimer's disease, the literature consistently reports a decrease in signal power in the alpha (8 13 Hz) and beta (13 30 Hz) ranges, increased theta (4 8 Hz) and delta (0.5 4 Hz) activity [3], and increased synchronization between certain regions of the brain, especially in low-frequency ranges. In frontotemporal dementia, deviations in the frequency organization of EEG signals are also observed, but these changes are localized more pronouncedly in the prefrontal area. A promising result from the implementation of machine learning approaches to solve dementia recognition tasks encourages research in this direction [4–8]. The authors of [9] conclude that the findings consistently suggest that frequency based multilayer resting state networks can be utilized to classify Alzheimer and frontotemporal dementia and have potential applications for clinical diagnosis. By extracting representative features, they apply them to 12 different classification algorithms. In other research [1] classification algorithms were also applied for dementia recognition. An unsupervised approach solving the problem of identifying cognitive states was applied in [4], but its use is limited. The advantage of such a search is the ability to find the natural structure of the studied data and on this basis to draw conclusions about the possibility of distinguishing cognitive groups.

The aim of the present study is to analyze and apply a pipeline of algorithms for processing EEG data to establish an effective procedure serving as a reliable tool for distinguishing patients' groups with different cognitive status such as Alzheimer's disease and frontotemporal dementia, from healthy people.

2 Materials and Methods

2.1 Data Used

The EEG data set used are time series recordings of the electrical activity of the brain in resting state that have been open data provided at [10] and described in detail in [1]. The recordings were made with 19 electrodes placed according to the international 10 - 20 system, with two reference electrodes A1 and $A2$. The electrodes covered the following scalp areas: prefrontal (Fp1, Fp2), frontal (F7, F3, Fz, F4, F8), temporal (T3, T4, T5, T6), central (C3, Cz, C4), parietal (P3, Pz, P4) and occipital (O1, O2) (Fig. 1). The duration of the participants' recordings is 12–14 minutes with a sampling rate of 500 Hz. From a total of 88 participants 36 were diagnosed with Alzheimer's disease (AD), 23 with frontotemporal dementia (FTD) and the remaining 29 were healthy controls (CN). The cognitive status of the participants was assessed according to the Mini-Mental State Examination (MMSE), with the mean score for the AD group being 17.75, for FTD 22.17 and for CN being 30.

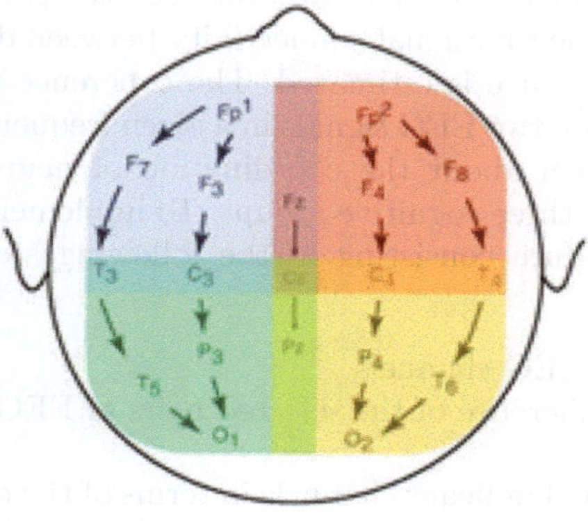

Fig. 1. Bipolar assembly with 19 EEG electrodes [8].

For the purposes of analysis, the raw data were preprocessed by applying several basic steps for signal purification [1]:

· Bandpass filtering in the frequency range 0,545 Hz using a Butterworth filter
· Recalculating the signals relative to the two reference signals A1 and $A2$.
· Detection and removal of artifacts such as muscle activity, body movements, electrical interference.
· Decomposition of EEG signals into several independent components, so that artifacts can be easily removed without losing brain information.
· Automatic classification and removal of artifacts. Components recognized as artifacts were removed, while components containing brain activity were preserved.

2.2 Methodology for Extracting Cognitive Groups in EEG Data

Selecting appropriate features of the EEG signals is a key objective to ensure the effectiveness of the analytical model for cognitive groups description. However, the task of extracting relevant features from EEG data that have the greatest discriminative ability to assess neurological status is difficult due to the high variability of EEG data. They are time-dependent and varying across brain channels, patients and clinical state.

One promising method for performing this task is based on assessing signal coherence as a measure of functional connectivity between different brain regions. Such research was successfully applied for identification of Alzheimer's and Mild

Cognitive Impairment patients [4]. In the present study, this approach is adapted and expanded with a focus on the following three groups' recognition: two groups with cognitive diseases - Alzheimer's and frontotemporal dementia, and a group of healthy persons. The research aims to investigate which combinations of the EEG signals of the brain regions in the distinct frequency bands and by what algorithm pipeline lead to best separation between the three groups.

In order to assess the functional connectivity between different brain regions the brain channel coherence is estimated. The coherence reflects the degree of synchronization between two EEG signals in a given frequency range. This allows conclusions to be drawn about the coordination of neuronal activity and by that to distinguish the three cognitive groups. To implement the above scenario, a data analysis procedure consisting of the following steps is applied to the preprocessed data:

· Selecting pairs of EEG channels
· Calculating the coherence of the selected pairs of EEG channels in different frequency ranges
· Choosing the most significant channels in terms of the coherence dependency
· Applying the discriminant analysis for data space reduction
· Clustering to obtain data groups
· Evaluation of the resulting groups in terms of recognition of patients' diagnoses.

3 Data Analysis Procedure

In this section, the procedure steps are described in detail.

3.1 Selecting of EEG Channels Pairs

This stage looks at specific combinations of pair channels that could further assess the connection of the different areas of the brain. The goal is to determine how different areas of the brain interact with each other. For example, combinations of channels located in the frontal and temporal regions can reflect interactions related to cognitive processes such as attention and memory, while channels located in the parietal and occipital regions can be used to study visual processing or spatial orientation.

The selection of appropriate channels is extremely important because each combination can provide unique information about brain activity and therefore potential pathologies. Channel pairs can be selected based on the known anatomical and functional relationships or on the hypotheses of interaction related to research on a given condition. For example, in patients with Alzheimer's disease, reduced coherence can be observed in combinations of channels related to cognitive functions, which is an indicator of impaired communication between brain regions that is the result of a loss of structural and functional integrity of the long corticocortical tracts [2]. Pairs such as P3O1 and Fp1F3 or P4O2 and Fp2F4 reflect interactions between the parietal-occipital and frontal cortex, which play a key role in sensory information processing and cognitive control. The channel pairing originally introduced in the Alzheimer and Mild Cognitive Impairment discrimination task [4] is realized for the assembly of 16 channels. We adopt this approach but extended it with the central EEG electrodes Fz, Cz, Pz due to 19 channels assembly in our case. Thus, the corresponding two pairs Fz-Cz and Cz-Pz (Fig. 1) have been included for further feature selection analysis.

3.2 Coherence Calculation in EEG Signals

Assessment of the channel's coherence aims to reveal if channel dependency exists. Coherence $Cxy(f)$ is a normalized value that indicates the degree of dependence between two signals $x(t)$ and $y(t)$ at a certain frequency f:

$$Cxy(f) = \frac{|Pxy(f)|^2}{Pxx(f)Pyy(f)}, \tag{1}$$

where $Pxy(f)$ is the cross-spectral density of the signals $x(t)$ and $y(t)$, t denotes the time and $Pxx(f)$ and $Pyy(f)$ are the auto-spectral density of respectively the signals $x(t)$ and $y(t)$ at frequency f. The coherence values range from 0 - completely unrelated at this frequency and to 1 full linear dependency at the considered frequency. If a strong linear relation is observed, this may indicate full synchronization between the two signals in the considered frequency.

The coherence is estimated for different frequencies that are of particular interest for signal distinguishing. These are both standard frequency bands, and further additional subdivision of some frequency bands could be reached as a better signal division of the cognitive disorders. The five standard frequency bands [1,13,14] are investigated as they provide information on the intensity of neural activity and could serve as markers of neurophysiological functioning. They are:

· Delta (δ) 0.5 4 Hz frequency is low-frequency activity that is often associated with sleep, deep relaxation, or pathological states. Patients with neurodegenerative diseases may exhibit increased power in this band.
· Theta (θ) 4 8 Hz frequency is associated with dreaming, meditation, and creative thinking. Increased theta activity in the resting state is characteristic of cognitive deficits, including dementia.
· Alpha (α) 8 13 Hz frequency represents rhythmic activity in a calm, awake state with eyes closed. Often, it is an indicator of functional brain maturity and cognitive integration. Reduced α-power is characteristic of patients with Alzheimer's disease.
· Beta (β) 13 30 Hz frequency is associated with active thinking, concentration, problem-solving, and anxiety. Patients with neurodegenerative diseases often exhibit alterations in this band, including both decreased and increased power depending on the stage and type of disease.

· Gamma (γ) 30 40 Hz frequency is associated with higher cognitive functions, including attention, memory, sensory integration, and conscious processing. Gamma activity is often used as an indicator of neural coordination between different brain regions.

Investigation of the coherency in additional division of the alpha and beta frequencies bands is reasonable due to existing knowledge of the sensitivity of brain activity in these bands in cases of cognitive diseases [4, 13].

3.3 Feature Selection

EEG data are noisy and heterogeneous, which requires careful research to assess their significance in terms of channels' coherence. Here, the channel' dependency is estimated by the Wilcoxon test that enables statistical selection of those channels that are at most dependent, and thus to serve as significant features for group discrimination. The Wilcoxon rank-sum test is used as a nonparametric method to compare two independent samples. The test does not require normality of the distribution, making it particularly suitable for EEG data that often present strong outliers and artifacts. Furthermore, by applying a significance level of 0.03 instead of the standard 0.05, a more stringent selection criterion is set. It aims to detect truly robust and meaningful differences that are not the result of random noise.

3.4 Multiple Discriminant Analysis

Following the methodology, the number of features identified is large as the coherence dependency is estimated for all received channel pairs and for all frequencies of interest. After statistical selection the number of features still remains large. This makes the discriminant task difficult being problematic and inefficient. In order to deal with this Multiple Discriminant Analysis (MDA) is applied in the next step of the data analysis. It reduces the data dimensionality and visually distinguishes the groups found. MDA is a linear dimensionality reduction method that simultaneously optimizes the separability between predefined classes. It seeks to find linear combinations of features that maximize the differences between classes while minimizing the scatter within classes.

MDA is applied on the extracted features EEG channels, which average be-tween pairs coherence in the different frequency bands is over significance level of 0.03. Since our task involves multigroup classification due to presence of three groups Alzheimer's disease patients, frontotemporal dementia patients and healthy con-trols, MDA uses reduction to two discriminant components (MD1 and MD2) that maximally separate the available groups.

Before applying MDA, the features data were standardized, which is especially important to reduce the sensitivity to the scale of the features.

3.5 Identification of Cognitive Groups

Nevertheless, that there is information about the existing groups of participants our goal was first to reveal the natural data division and then to interpret the result according to the participants' diagnoses. Thus, in the final stage of the analysis, a method for unsupervised learning is applied. It is based on clustering with the k-means algorithm. The main goal is to group the participants into clusters that best reflect the hidden structure in the data. The algorithm aims to minimize the Within cluster sum of squares of the data to the cluster centers and by that optimizing the structure of the data relative to the clusters prototypes to which they belong. Euclidean distance is applied that selects spherical clusters. It is an acceptable assumption as we expect to identify groups of participants with similar characteristics of cognitive disorders in each group that are equally spread around the particular cluster prototype value. The clustering by k-means is a suitable choice in cases where simple but effective segmentation is sought. It is also a competing method with simpler implementation and robust behavior.

The proper number of clusters was found according to Elbow Method. Further, we are assessing to what extend the found clusters correspond to the three participants groups known by their diagnosis: Alzheimer's disease, frontotemporal dementia and healthy control. For this aim three different indexes have been used to reveal different characteristics of the comparison between diagnosis groups and clusters found.

Adjusted Rand Index (ARI) is a metric measure for assessing the similarity between real and the estimated clusters, which accounts for and corrects for the influence of random coincidence [11]. It normalizes the Rand Index by the expected value E under random clustering:

$$ARI = \frac{R - E}{max(R) - E}, \tag{2}$$

where R denotes Rand Index - a basic measure used to compare two clustering results based on the matches between all possible pairs of objects. R measures the relative number of pairs of objects that are classified the same way in both groupings - either both in the same cluster or both in different ones. ARI is bounded in the interval $[-1, 1]$, with a value of 1 indicating perfect agreement between the two clustering solutions, a value of 0 corresponding to the result under random distribution, and negative values indicating a worse case than random distribution.

Normalized Mutual Information (NMI) is a metric that compares the two groupings - first the known classification of disease states as Alzheimer's disease, frontotemporal dementia and the control group and the second one obtained by the clustering algorithm. NMI measures how much information about the one distribution can be extracted from the other. NMI is calculated as a normalized version of the mutual information between the two groupings [12]. As much is the mutual information then the two groupings are more overlapped. It is a value between 0 for low and 1 for full overlapping.

The index CR measures the proportion of elements in a set of data that, based on the clustering results, are correctly assigned to their corresponding real classes:

$$CR = \frac{1}{n} \sum_{i=1}^{c} \max_{j} |C_i \cap L_j|, \tag{3}$$

where n is the number of data being clustered, c is the number of clusters, C_i is the set of data assigned to the i-th cluster, L_j is the set of data belonging to the j-th known group, and $\max_j |C_i \cap L_j|$ is the number of objects in the cluster i that mostly match the label j.

In addition, the classification accuracy is considered separately for each of the three studied groups to assess how accurately the clustering reproduces the known grouping of data for each specific diagnosis. For each group G, where G stands for group of AD, FTD or CN respectively, the group classification accuracy CR_G is calculated by the following formula:

$$CR_G = \frac{1}{|G|} \max_{k} |G \cap C_k|, \tag{4}$$

where $|G|$ is the number of objects in the respective group G, C_k is cluster k, and $|G \cap C_k|$ is the number of data points belong to group G and fall into cluster C_k.

4 Experimental Results and Analysis

The selection of channel pairs reflects anatomical and functional connections between key areas of the cerebral cortex involved in various important cognitive processes. The channel pairing organized for further analysis is shown in Table 1. For this we follow the pairing introduced for bipolar assembly of 16 electrodes [4], where each pair is numbered. This pairing is extended by the central electrodes' pairs (Fz, Cz) and (Cz, Pz) with their respective number 17 and 18. The last column of the table named "Received Pair Number" consists of numbers that are concatenations of the numbers of the two channel pairs given at the respective row. The channels' pairing marked in this column are further investigated for significance of the signal coherence.

Different experiments have been organized in searching for the significant features. The coherence of each selected channel pair was computed by eq. (1) estimated for the mean value in the respective frequency bands. The mean coherence value of the organized pairs is finally selected according to the Wilcoxon rank-sum test.

The results obtained for the five standard frequency bands are investigated and presented at Table 2. There are channel pairs that are statistically significant in only one of the group comparisons at the present frequency, but do not repeat in the others. For example, a given pair may appear significant in the AD vs. CN comparison as 19 of delta band and 19 and 513 of theta band, but be absent

Table 1. Channel pairing for bipolar mounting.

Channel Pair Electrodes	Pair Number	Channel Pair Electrodes	Pair Number	Channel Pairs	Received Pair Number
P3-O1	4	F3-C3	2	P3-O1 - F3-C3	42
P3-O1	4	Fp1-F3	1	P3-O1 - Fp1-F3	41
P3-O1	4	Fp1-F7	9	P3-O1 - Fp1-F7	49
P4-O2	8	Fp2-F4	5	P4-O2 - Fp2-F4	85
P4-O2	8	Fp2-F8	13	P4-O2 - Fp2-F8	813
P4-O2	8	F4-C4	6	P4-O2 - F4-C4	86
Fp1-F3	1	C3-P3	3	Fp1-F3 - C3-P3	13
Fp1-F3	1	T5-O1	12	Fp1-F3 - T5-O1	112
Fp2-F4	5	C4-P4	7	Fp2-F4 - C4-P4	57
Fp2-F4	5	T6-O2	16	Fp2-F4 - T6-O2	516
P3-O1	4	C3-P3	3	P3-O1 - C3-P3	43
P3-O1	4	T3-T5	11	P3-O1 - T3-T5	411
P4-O2	8	C4-P4	7	P4-O2 - C4-P4	87
P4-O2	8	T4-T6	15	P4-O2 - T4-T6	815
Fp1-F3	1	F3-C3	2	Fp1-F3 - F3-C3	12
Fp1-F3	1	Fp1-F7	9	Fp1-F3 - Fp1-F7	19
Fp2-F4	5	F4-C4	6	Fp2-F4 - F4-C4	56
Fp2-F4	5	Fp2-F8	13	Fp2-F4 - Fp2-F8	513
Fz-Cz	17	Cz-Pz	18	Fz-Cz - Cz-Pz	1718

Table 2. Significant channel pairs detected by Wilcoxon test of the 5 main frequency bands.

Groups	Frequency bands				
	δ	θ	α	β	γ
AD vs. CN	19	513	411	411	
	1718	19	86	1718	
			85	42	
			12	86	
			1718	43	
			43	815	
			42	87	
			13		
			41		
			815		
			57		
			87		
FTD vs. CN			411	1718	1718
			86	43	
			1718	815	
			42	87	
			43		
			815		
			87		
AD vs. FTD		42			

in both FTD vs. CN and AD vs. FTD comparisons. Such non-repeating pairs can be interpreted as potentially biomarkers, characteristic for distinguishing the two particular groups for a specified frequency.

The result of Table 2 clearly shows that the alpha and beta bands are more sensitive than the others, as they expose more channels that are significant for distinguishing the cognitive groups. Based on this and due to the certain knowledge about dependency between dementia diseases and the α and β bands activities [4,13] additional division of these two bands are important to investigate. These subbands are low α range (6 - 9 Hz) and high α range (9 - 13 Hz) as well as low β range (13 16 Hz), mid β range (16 20 Hz) and high β range (20 24 Hz). It increases the sensitivity of the analysis to the neuropathological processes as new channel groups are recognized Tables 3 and 4.

Table 3. Significant channel pairs detected by the Wilcoxon test if the frequency α is divided into two subgroups.

Compared groups	Frequency bands					
	δ	θ	low α	high α	β	γ
AD vs. CN	19	49	513	43	411	
	1718	12	19	86	1718	
		13		87	42	
		57		41	86	
		513		1718	43	
		1718		85	815	
		19		57	87	
		516		411		
		56		42		
				13		
				12		
				815		
FTD vs. CN			43	43	1718	1718
				86	43	
				87	815	
				1718	87	
				411		
				42		
				815		
AD vs. FTD		43				
		42				
		1718				

Table 4. Significant channel pairs detected by the Wilcoxon test if the β frequency is divided into three subgroups.

Compared groups	Frequency bands					
	δ	θ	α	low β	mid β	high β
AD vs. CN	19	513	1718	86	815	43
	1718	19	13	411	42	87
		56	86	42	1718	42
			87	41	41	411
			57	43	43	1718
			85	815	411	86
			815	87	86	815
			12		813	
			411		87	
			43			
			42			
			41			
FTD vs. CN			1718	87	815	43
			86	43	42	1718
			87	815	1718	86
			815		43	815
			411		86	
			43		87	
			42			
AD vs. FTD		42				

Difference between AD and FTD participants is only sensitive for the connection P3-O1 - F3-C3 (pair 42) in theta band (Tables 2 and 4). These connections are sensitive for alpha band division as well (Table 3) but added with connections P3-O1 - C3-P3 coded by 43 and the central channels Fz-Cz - Cz-Pz coded as 1718. These results underline the role of the brain left side activity enabling us to distinguish the two dementia.

In order to reduce the data dimensionality and by that to improve the structuring ability of the clustering the method of Multiple Discriminant Analysis was applied on the three spaces formed by respective significant features (Tables 2, 3, 4). As a result, three two-dimensional data spaces are formed, which after that are clustered.

An important task of the unsupervised data structuring and in particular applying the k-means algorithm is to investigate the appropriate number of clusters in the data space. We deal with this problem by an assessment of the Elbow Method that calculates the sum of the squared distances (SSD) between the data and corresponding cluster center estimated for the data partition from 2

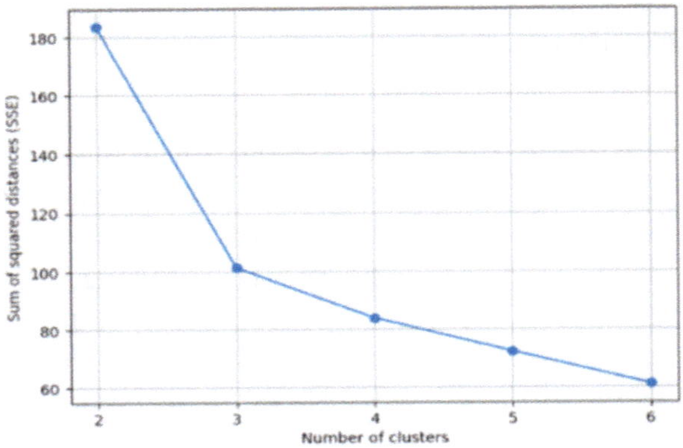

Fig. 2. SSE values in respect to the number of clusters of the five main frequencies bands data

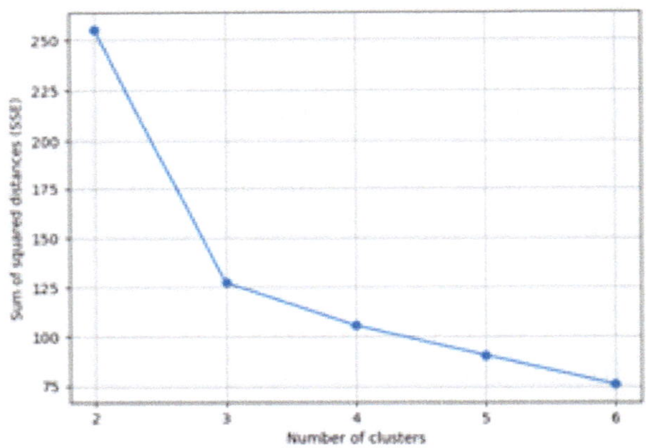

Fig. 3. SSE values in respect to the number of clusters of the data with two subbands of the alpha frequency

to 6 clusters. Figures 2, 3 and 4 show the resulting curves of the correspondingly three studied data spaces. It clearly could be seen and thus to conclude that division in three clusters is most acceptable for all data spaces. By that, the clustering in three clusters is the next stage of the data analysis procedure.

The other important task of the procedure is to find a proper initialization of the k-means algorithm i.e. the initial cluster centers values. In case of pure unsupervised data structure identification, the random initialization with k-means++ is a commonly accepted strategy. Ten-fold clustering was applied and the average value of the clustering indexes for assessment of the grouping in the three data spaces were calculated (Table 5).

The data set of statistically significant features obtained by division of beta frequency into three subbands leads to the highest results in the clustering task.

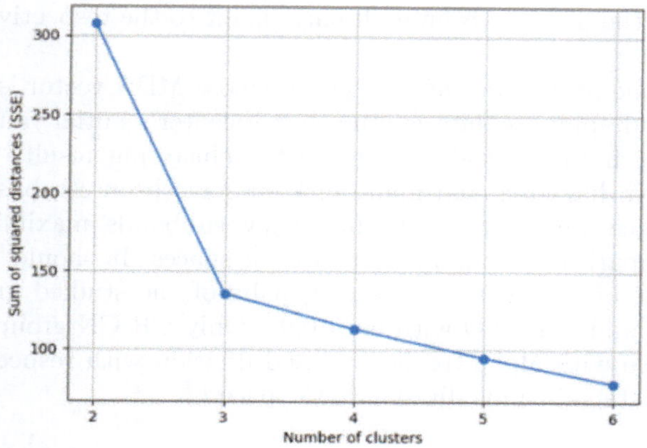

Fig. 4. SSE values in respect to the number of clusters of the data with three subbands of the beta frequency

The division of the beta frequency allows for more precise differentiation of cognitive states by capturing subtle spectral differences specific to each subgroup. Including two alpha frequency bands as separate features has a positive impact on the quality of clustering as for this case the best separation of FTD against the controls is ensured with level 0.67. The procedure results show good distinguishing between groups, with a clear performance for measuring the proportion of elements CN having values above 0.65, whereas the metric measure for assessing the similarity between real and the estimated clusters by index *ARI* is not sensitive for this task.

Table 5. Clustering indexes of clustering with random initialization.

Cluster index	5 main frequency bands	With two subbands of α frequency	With three subbands of β frequency
ARI	0.0997	0.0346	0.1754
NMI	0.3798	0.3378	0.4493
CR	0.6083	0.5847	0.6792
CR - AD	0.62	0.54	0.68
CR - CN	0.67	0.55	0.70
CR - FTD	0.47	0.67	0.62

In looking for better clustering we investigate another approach as well. Initial values of the cluster centers were set by a semi-unsupervised clustering. Relying on the fact that three clusters are searching for, the following steps were incorporated in the procedure:

1) Take all data of the MDA data space of each of the known diagnosis group AD, FTD and controls.

2) Calculate the mean vector of each group.

3) Measure the distance from each participant to the respective mean vector of its group.

4) Choose one participant of each group whose MDA vector is closest to the mean vector and apply its data for the cluster center's initial values.

This approach substantially improves the clustering results as large overlapping between diagnoses and found clusters have been established (Table 6). Again, the space with three beta frequency subbands maximizes the three indexes in comparison to the other two data spaces. It should also underline the high degree of covering for assessing pairs of the studied groups: CR-AD with level 0.85 and CR-FTD with level 0.92. Only CR-CN group is better recognized in space with alpha frequency band division with respective level 0.90 than 0.85 for beta subbands division data space.

Table 6. Clustering indexes of semi-unsupervised clustering.

Cluster index	5 main frequency bands	With two subbands of α frequency	With three subbands of β frequency
ARI	0.3695	0.5693	0.6133
NMI	0.5619	0.7321	0.7559
CR	0.7403	0.8514	0.8736
CR - AD	0.70	0.77	0.85
CR - CN	0.82	0.90	0.85
CR - FTD	0.65	0.90	0.92

5 Conclusion

The paper presents a procedure for distinguishing three groups of participants those with Alzheimer disease, with frontotemporal dementia, and healthy people, based on their EEG data in resting state. The procedure relays on rigorous feature selection and follows clustering analysis in order to define the groups.

The use of statistically significant features defined by channel pair coherence leads to a substantial improvement in the quality of clustering compared to using all available variables from the original dataset. Increased values of the clustering assessment metrics demonstrate that informed feature selection is critical for successful modeling of EEG data. It should be underlined that significant features for discrimination of frontotemporal dementia were discovered but with reduced amounts in comparison the other two groups. The division of the beta frequency band into three subbands - low, medium and high range, allows for a more precise accounting of neurophysiological differences between groups. Additional division of the alpha band also increases the ability to group compared to using only the five standard frequency bands. Inclusion of the central EEG electrodes (Cz, Pz, Fz) contributes to a better performance of the clustering model, especially in the classification of healthy individuals and Alzheimer's patients.

A significant improvement in the ability to find the three groups was achieved using a semi-supervised learning approach by identifying good initial values of the cluster centers that are close to their extreme values.

Acknowledgments. This research is part of the GATE project and funded by the Horizon 2020 WIDESPREAD-2018-2020 TEAMING Phase 2 programme under grant agreement no. 857155, the programme "Research, Innovation and Digitalization for Smart Transformation" 2021-2027 (PRIDST) under grant agreement no. BG16RFPR002-1.014-0010-C01 and the project SUMMIT BG-RRP-2.004-0008-C01.

References

1. Miltiadous, A., et al.: A dataset of scalp EEG recordings of Alzheimer's disease, frontotemporal dementia and healthy subjects from routine EEG. Data **8**(6), 95 (2023)
2. Leuchter, Andrew F., et al.: Changes in brain functional connectivity in Alzheimer-type and multi-infarct dementia. Brain **115**(5), 1543–1561 (1992)
3. Jeong, J.: EEG dynamics in patients with Alzheimer's disease. Clin. Neurophysiol. **115**(7), 1490–1505 (2004)
4. Akrofi, K., Baker, M., O'Boyle, M., Schiffer, R.: Clustering and modeling of EEG coherence features of Alzheimer's and mild cognitive impairment patients. In: Proceedings of the Annual International Conference of the IEEE Engineering in Medicine and Biology Society , vol. 2008, pp. 1092–1095 (2008)
5. Aljalal, M., Aldosari, S., Molinas, M., Alturki, F.: Selecting EEG channels and features using multi-objective optimization for accurate MCI detection: validation using leave-one-subject-out strategy. Sci. Rep. **14** (2024)
6. Baker, M., Akrofi, K., Schiffer, R., O'Boyle, M.: EEG patterns in mild cognitive impairment (MCI) patients. Open Neuroimaging J. **2**, 52–55 (2008)
7. Hegde, N.N., Nagananda, M.S., Harsha, H.: EEG signal classification using K-Means and Fuzzy C means clustering methods. Int. J. Sci. Technol. Eng. (IJSTE) **2**(1), 83–87 (2015)
8. Matos, J., Peralta, G., Heyse, J., Menetre, E., Seeck, M., van Mierlo, P.: Diagnosis of epilepsy with functional connectivity in EEG after a suspected first seizure. Bioengineering **9**(11), 690 (2022)
9. Yajing, S., et al.: Differentiating between Alzheimer's disease and frontotemporal dementia based on the resting-state multilayer EEG network. IEEE Trans. Neural Syst. Rehab. Eng. **31**(1), 4521–4527 (2023)
10. Taghia, Y.: Open Nuro Dataset, Kaggle (2022). https://www.kaggle.com/datasets/yosftag/open-nuro-dataset/data
11. Sabu, S.: Rand index (RI) vs. adjusted rand index (ARI) in K-means clustering. Medium **23** (2024)
12. Mahmoudi, A., Jemielniak, D.: Proof of biased behavior of Normalized Mutual Information. Sci. Rep. **14**(1), 9021 (2024)
13. Nayak, C.S., Anilkumar, A.C.: "EEG normal waveforms" [Updated 2025 Aug 3]. In: StatPearls [Internet]. Treasure Island (FL): StatPearls Publishing, National Library of Medicine (2025)
14. Adebisi, A.T., Veluvolu, K.C.: Brain network analysis for the discrimination of dementia disorders using electrophysiology signals: a systematic review. Front. Aging Neurosci. **15**, 1039496 (2023)

Flare: An Efficient Spatial to Frequency Domain DNN for JPEG Artifact Correction

Stefan Petrov(✉) and Ivan Koychev

Faculty of Mathematics and Informatics, Sofia University St. Kliment Ohridski, Sofia, Bulgaria
{sgeorgievp,koychev}@fmi.uni-sofia.bg

Abstract. JPEG is a widely used compression method with a known drawback that it leads to the appearance of block artifacts when compressing with low quality factor. Block artifacts are tonal discontinuities between adjacent blocks of 8×8 pixels. In this work, we propose a new model called Flare to remove these block artifacts from singly-compressed JPEG images. Flare is a function from the spatial to the frequency domain of the image and removes artifacts in both monochrome and color images without any limitation in resolution or quality factor with which the image is compressed. Compared to previous works, our model shows near state-of-the-art results in PSNR and state-of-the-art in SSIM (although differences are imperceptible to the human eye).

Keywords: image processing · JPEG artifacts · Discrete cosine transform · image compression · deep learning

1 Introduction

JPEG is still the most widely used compression method for still images. It's in the class of lossy compression techniques and uses the Discrete cosine transform (DCT) to represent the image in the frequency domain. The DCT coefficients are further quantized. The level of compression is controlled by a **quality** parameter in the encoder (q), which takes values in the range [1, 100]. The encoder starts by dividing the image into 8×8 pixel blocks and then compresses each block individually. This means that the DCT and quantization are applied per block. For values of $\mathsf{q} \in [1, 50]$, the quantization is so strong that discontinuities start to appear at block boundaries. These discontinuities are known as **block artifacts**.

Restoring information lost due to compression is a long-studied topic in image processing. Early methods did not try to restore it, but rather mitigate the visual effect of it. However, this resulted in loss of details, hence more loss of information. The recent deep learning revolution inspired many researchers to investigate deep learning techniques for image restoration.

In this paper, we present a new deep neural network (DNN) with a very simple architecture and an option for manual tuning that achieves impressive

I. K. Koychev (Ed.): ERIS 2025, PINNS 4, pp. 73–89, 2026.
https://doi.org/10.1007/978-3-032-30153-6_5

results for JPEG deblocking using only a fraction of the parameters of FBCNN [7]. Our network can be viewed as a function from the spatial to the frequency domain. Its design is closely intertwined with the algorithms of encoding and decoding JPEG images, and it directly estimates the residuals of the DCT coefficients that were lost in the quantization step during encoding (compression). We also introduce a novel and better sampling approach for q that is used for data generation during training. We found this approach (see Sect. 4.1) to speed up convergence during training, so future works can benefit from it.

This paper proceeds as follows: Sect. 2 gives an overview of the current methods for de-blocking. Section 3 introduces the general architecture of the proposed model, and Sects. 3.1 and 3.2 present the details. Section 4 provides a brief overview of dataset generation, and Sect. 4.1 introduces the novel q-sampling. Section 5 describes the training details. In Sect. 6 we present the results and compare them to existing works.

2 Related Work

Existing methods for the reduction of JPEG block artifacts can be broadly summarized into two categories: filtering and reconstructing. The filtering ones try to relax the visual discontinuities at block boundaries. The reconstructive ones try to recover the lost information - they are defined as an optimization problem of minimizing the distance between the compressed image I_c and the original image I_o. One of the first works on filtering methods comes from the domain of video compression [14]. In it the authors describe an adaptive deblocking filter used in the H264/MPEG-4 video coding standards. The proposed filter is adaptive, and its strength depends on the context of the current frame - areas with more detail get less deblocking, areas with less get more deblocking. [12] proposes a filtering algorithm based on wavelets. Perhaps the most successful work in the filtering category is Shape-adaptive DCT [6]. This algorithm is a modified version of the DCT, initially developed for noise reduction [5]. Its idea is to apply DCT on segments depending on the shape of the object (or patterns) in the image, rather than square blocks. That way, image structure is better represented, which makes noise suppression and block removal more efficient and detail preserving. In [24], a technique based on AdaBoost is presented, which offers better performance than Shape-adaptive DCT. All filtering methods have the common shortcoming that they reduce the sharpness of the image and produce images with smoothed textures.

Reconstructive methods assume that compression is an image degradation process and aim to reduce it by reconstructing the non-degraded image. Reconstructive methods for JPEG artifacts correction can be divided into two categories: reconstructive for single q and reconstructive for range of q. The reconstructive for single q are optimized for the removal of block artifacts only from images compressed with a particular quality factor q, hence their applicability is limited. Early examples of reconstructive methods with single q are the POCS-based method in [26], Fields of Expert Priors [23], Sparse Coding Representation [8], semi-local Gaussian Processes-based solution [10], and A+ [22].

In [16], the authors apply sparse coding in the frequency domain to eliminate the quantization errors. Building on top of this idea, [25] replace the sparse coding step with a deep neural network and achieve better results.

ARCNN is the first de-artifacting method based on a convolutional neural network (CNN) [3] [27], although results are reported only for grayscale images. DnCNN [28] uses a deeper convolutional network with residual connections and significantly outperforms ARCNN. DPW-SDNet [2] combines wavelet transformation with CNN. The architecture consists of two parallel branches - P-SDNet and W-SDNet. P-SDNet reconstructs the image in the spatial domain, W-SDNet in the wavelet domain. The outputs of both branches are concatenated and projected into a single channel. DmCNN [29] uses a similar architecture, but instead of a branch in the wavelet domain, uses a branch in the frequency domain, which reconstructs the DCT coefficients. DmCNN's output is a weighted sum of the input image and the outputs of both branches. MWCNN [15] combines the U-Net architecture with multi-level wavelet transformation instead of dilated convolution. MWCNN is reported to have better computational efficiency and accuracy than DnCNN. IDCNN [30] is the first model trained on color images. IDCNN again uses information from the frequency and the spatial domain to restore the image. One of the more recent examples for single-q reconstruction is SwinIR [11]. SwinIR is a modified version of the Swin Transformer [17] for the goals of JPEG artifact reduction and achieves state-of-the-art results for grayscale images.

All of the above methods require training different model for a particular q. Now we turn our attention to models able to reconstruct images for range of q-s. In [31], the authors propose an autoencoder architecture with RRDB blocks to remove block artifacts. The proposed model enhances the Y, Cb, and Cr planes in 2 steps. First, the Y plane is enhanced, and then the Cb and Cr planes are enhanced conditionally over the enhanced Y plane from step 1. RRDB demonstrates flexibility, although it is trained on a mix of compressed images with q ranging from 10 to 100. AGARNet [9] uses a local q estimator Q_{EST} which partitions the image into blocks, and compression quality q is estimated for each block. After that, the reconstruction process is conditioned on the estimated quality. Reconstruction is again done in the spatial and frequency domains. In [4], instead of a dedicated q estimator, the authors use the quantization tables, written in the JFIF[1] file to learn q-independent representation for each block 8×8 of the image. The representation is then passed to a deep CNN to reconstruct the final image. Here, the reconstruction process is entirely in the frequency domain and consists of two steps: first, the Y plane is reconstructed, and then the Cb and Cr planes are conditionally reconstructed from it. The state-of-the-art model FBCNN [7] is based on U-Net [21] and features a q estimator at the bottleneck, and it is the first model trained to enhance doubly compressed

[1] Technically, the file format used to store JPEG encoded data is JFIF (JPEG File Interchange Format). JFIF adds essential header information for cross-platform compatibility.

images. FBCNN also allows manual tuning of the strength of reconstruction in the decoder.

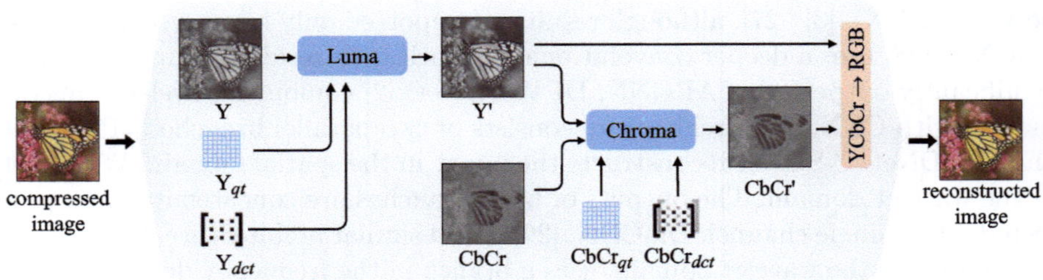

Fig. 1. Overview of Flare's Image Reconstruction Pipeline. Reconstruction is divided into 2 phases: luminance and chrominance. The blue rounded rectangles are the Flare Luma and Flare Chroma subnets. First, the luminance plane Y is reconstructed with Flare Luma (denoted Y′). Then Y′ is concatenated with CbCr and passed to Flare Chroma to reconstruct the chrominance. Denote the reconstructed chrominance with CbCr′. Finally, Y′ and CbCr′ are concatenated, and the image is converted to RGB.

3 Proposed Model Architecture

In this section, we describe the architecture of the proposed model **Flare**. Flare is a deep convolutional network consisting of luminance reconstructor (Y plane) **Flare Luma** and chrominance reconstructor (CbCr planes) **Flare Chroma**. These reconstructors are learned functions from the spatial to the frequency domain that directly estimate the quantization error Δ_q:

$$f : I_C \rightarrow \Delta q = \mathrm{DCT}(I_O) - \mathrm{DCT}(I_C), \tag{1}$$

where I_C is the image after compression, I_O is the original image (before compression) and DCT is the Discrete cosine transform.

While most existing deblocking model architectures are general enough to be used in denoising and superresolution tasks, Flare's architecture is specialized only for JPEG de-blocking. The architecture is tailored specifically to the JPEG encoder - it is conditioned on the quantization tables stored in the JFIF file to eliminate the artifacts. The process to remove block artifacts from an input image I_C is illustrated in Fig. 1. The reason we choose to divide the reconstruction process into two stages is that the JPEG encoder often downsamples the Cb and Cr planes.[2] We'll denote the output from Flare Luma with Y′ - this is the reconstructed Y plane. Conditioning chrominance reconstruction on Y′ enables the restoration of details lost due to chroma downsampling in the final output

[2] The most common chroma downsampling mode is 4:2:0, in which the height and width of the chrominance are halved.

image. This architectural choice also allows Luma and Chroma subnets to be trained in isolation.

In the following subsections, we describe in more detail the architecture of the subnets.

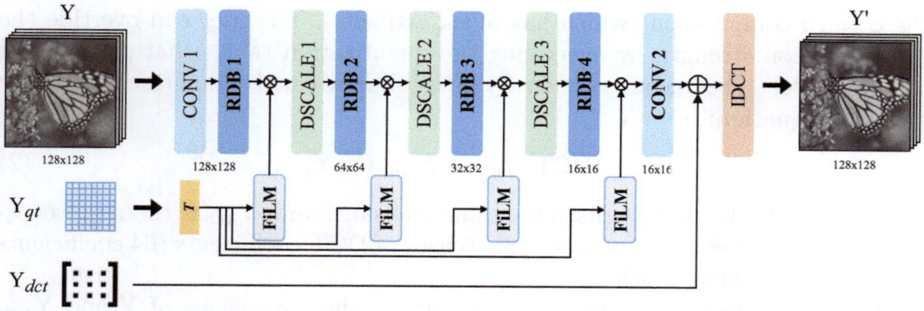

Fig. 2. Architecture of Flare Luma. The first layer in Flare Luma is convolutional, and it increases the number of channels from 1 to C, where C is a hyperparameter denoting the 'base number of channels'. Following the first layer are 4 Residual dense blocks (RDB). Leaky RELU is used as a nonlinearity in the RDB blocks. The output activation maps from each RDB block are modulated by FiLM, which is conditioned on the Y quantization table. The height and width of the input activation maps for each RDB block are written under it. The yellow rectangle T after Y_{qt} is a linear transformation $T : [1, 255] \to [0.0, 1.0]$. Between every pair of RDB blocks, there is a learned 2X convolutional downsampling that reduces the height and width and increases the number of channels by a factor $k = 2$. The last layer is again convolutional and projects the output activation map from the last RDB block to a tensor with dimensions $H/8 \times W/8 \times 64$ (equal to the dimensions in Y_{dct}). The computed residual $\Delta_{\mathbf{q}}^{Y}$ is added to Y_{dct}, and the result is converted to the spatial domain using the Inverse Discrete Cosine Transform (IDCT).

3.1 Architecture of Flare Luma

The architecture of Flare Luma is illustrated in Fig. 2. Flare Luma removes the block artifacts from the luminance plane in the image. The network estimates the residual Δ_q^Y lost due to quantization and adds it to the DCT coefficients of the input Y plane. The backbone of Flare Luma consists of 4 Residual dense blocks (RDB) each followed by learned 2X convolutional downscaling. To condition Δ_q^Y on q, we use FiLM [20] on the activation maps right before downscaling. The quantization table modulates FiLM for the Y plane (Y_{qt}). Since values in Y_{qt} are proportional to the strength of quantization and inversely proportional to q, using Y_{qt} as a modulator will result in small output residuals when q is large. For images compressed with $\mathsf{q} = 100$, Δ_q^Y will be 0. The FiLM mechanism enables Flare Luma to reconstruct images with arbitrary q (in the literature, this is known as 'blind reconstruction', because q is not explicit input to the model). FiLM also allows the user to control the strength of reconstruction by setting manually Y_{qt} or equivalently q. For example, if a JPEG image is double compressed (first with $q = 20$, then with $q = 80$), the quantization tables for $q = 20$ will be overwritten by those for $q = 80$, hence block artifacts won't be removed, because Flare will reconstruct the image that is the output from the $q = 20$ compression (which has block artifacts). The user can override the reconstruction strength by overriding the quantization tables that go as input to the model. The reconstruction by Flare Luma can be described with the following equation:

$$Y' = \text{IDCT}(Y_{dct} + f_{luma}(Y, Y_{qt})), \tag{2}$$

where IDCT is the Inverse discrete cosine transform and Y_{dct} is the representation of Y in the frequency domain - 3D tensor of DCT coefficients (64 coefficients for every 8×8 pixel block).

The downsampling stages are 3, because of the dimensions of Y and Y_{dct}. The height and width of the Y_{dct} tensor are 2^3 times smaller than the height and width of Y - each element in Y_{dct} corresponds to a single frequency in 8×8 block of pixels in Y.

3.2 Architecture of Flare Chroma

Flare Chroma has nearly identical architecture to Flare Luma. The main difference is that it conditions on the chrominance quantization tables (C_{qt}) and operates with 3-channel planes instead of 1. The inputs to Flare Chroma are Y', the compressed CbCr planes, and the quantization tables C_{qt}. The network outputs the quantization errors for the CbCr planes - Δ_{dct}^{Cb} and Δ_{dct}^{Cr}. These quantization errors (residuals) are transformed into a spatial domain using the IDCT and added to CbCr. The chrominance reconstruction process can be written as:

$$\text{CbCr}' = \text{IDCT}(\text{CbCr}_{dct} + f_{chroma}(\text{CbCr}, Y', C_{qt}) \tag{3}$$

Note that conditioning on Y' is done by concatenating Y' to CbCr, while conditioning on C_{qt} is done using FiLM. Adding Y' to the input allows the model to recover lost details in the chrominance planes due to subsampling. If chrominance is subsampled, then Y' is rescaled to the resolution of CbCr. We cannot upscale CbCr, because the network outputs residuals with the same shape as the Cb_{dct} and Cr_{dct}. We also experimented with an additional superresolution block for the reconstructed CbCr planes, but found no improvement in PSNR. Finally, the reconstructed CbCr$'$ are concatenated with Y' and converted to an RGB image, which is the final output of the network. The architecture is shown in Fig. 3.

4 Dataset Generation

We used the Flickr2K [13] and DIV2K [1] datasets for training (the validation subset of DIV2K is used only for validation). Both datasets contain high-quality

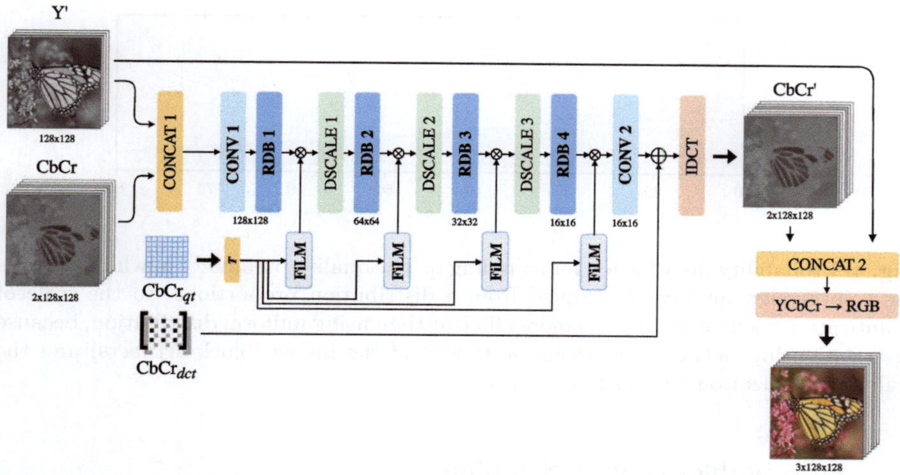

Fig. 3. Architecture of Flare Chroma. Flare Chroma shares the same backbone as Flare Luma - an initial convolution layer that projects the number of channels to C, and four consecutive RDB blocks with 2X downscaling between each pair. The backbone operates on all three planes, intermixing luminance with chrominance to add detail. Then convolution follows, which projects the final activation maps to a tensor with shape $H/8 \times W/8 \times 128$. This tensor is the Δ_{dct}^{CrCb} residuals that are added to CbCr$_{dct}$. Then IDCT converts chrominance to the spatial domain. Denote the reconstructed chrominance with CbCr$'$. Y$'$ and CbCr$'$ are concatenated and converted to RGB.

PNG images. From each image, we randomly crop a 128 × 128 patch. Larger patch sizes provide more context for reconstruction, but at the cost of slower training. We found that 128 × 128 is a good balance between context and train speed. To minimize disk IO, instead of cropping only one patch from an image, we crop 16 patches and add them to the current batch. We used a batch size = 32, so a single batch contains cropped patches from 2 images. Although this scheme may result in correlation between some of the examples in the batch theoretically (because they may overlap), in practice, the input images are so large relative to the patch that no overlap occurs. For validation, we use only central crops to keep metrics consistent - random crops will introduce additional noise in the metrics. From the cropped segments we prepare train examples (**x**, **y**) such that **x** is a JPEG compressed version of **y** with q $\in [10, 90]$.

For each input **x**, the planes Y, Cb, Cr, and their corresponding frequency representations Y$_{dct}$, Cb$_{dct}$, Cr$_{dct}$ are computed. We've implemented this computation in memory for performance reasons. Y, Cb, and Cr are scaled to [0,1]. Y$_{dct}$, Cb$_{dct}$, Cr$_{dct}$ are standartized with μ and σ calculated over all images in Div2K and Flickr2K per 8 × 8 block.

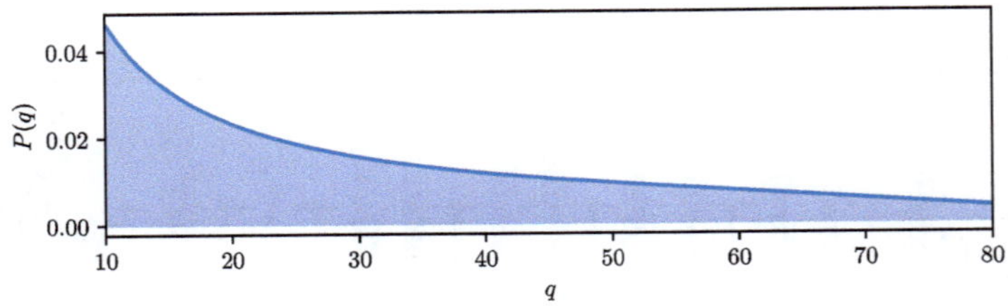

Fig. 4. Probability distribution for sampling q. The quality parameter q, which controls the compression quality, is sampled from a distribution proportional to the scale of quantization coefficients. This is more efficient than using uniform distribution, because the relationship between the visual aesthetics of the image (block artifacts) and the scale of quantization coefficients is linear.

4.1 Perceptable-Aware q Sampling

To generate $\mathbf{x}$, we need to sample a quality factor for the compression q. The naive approach is to use a uniform distribution $U(10, 90)$. But the relationship between q and the quantization coefficients is given by:

$$\mathbf{Q}_{ij} = \frac{s \times \mathbf{Q}_{ij}{}^{default} + 50}{100}$$

, where

$$s = \begin{cases} \frac{5000}{q} & q < 50 \\ 200 - 2q & q \geq 50 \end{cases}$$

Hence, for $\mathsf{q} \in [50, 100]$ is linear, but for $\mathsf{q} < 50$ the quantization coefficients grow progressively w.r.t. q. This non-linear scaling of the quantization coefficients makes block artifacts more prominent for small values of q. Since our goal is the reduction of block artifacts, we want the probability density for $P(\mathsf{q})$ to be representative of the visual presence of block artifacts. The visual presence is proportional to s, not $\mathbf{Q}$. That's why we use the probability density function $P(\mathsf{q}) \sim s$ (shown in Fig. 4). This choice of $P(\mathsf{q})$ increases the number of examples with prominent block artifacts in the train set, and experiments have shown that this speeds up convergence during training.

5 Training

Flare Luma and Flare Chroma are trained to minimize the *Charbonnier loss* given by:

$$\mathcal{L}(\mathbf{y}, \hat{\mathbf{y}}) = \frac{1}{B} \sum_{i=1}^{B} \sqrt{(\mathbf{y}_i - \hat{\mathbf{y}}_i)^2 + \epsilon}, \tag{4}$$

where ϵ is a small constant (in our case $\epsilon = 1 \times 10^{-8}$), and $\hat{\mathbf{y}}$ for Flare Luma and Flare Chroma are:

$$\hat{\mathbf{y}}^{\text{luma}} = \text{IDCT}(\mathbf{x}^{\text{Y}_{dct}} + f_{luma}(\mathbf{x}^{\text{Y}}, \mathbf{x}^{\text{Y}_{qt}})) \quad (5)$$

$$\hat{\mathbf{y}}^{\text{chroma}} = \mathcal{T}_{rgb}(\ \text{IDCT}(\mathbf{x}^{\text{CbCr}_{dct}} + f_{chroma}(\mathbf{x}^{\text{CbCr}}, \hat{\mathbf{y}}^{\text{luma}}, \mathbf{x}^{\text{C}_{qt}}))\ ), \quad (6)$$

where $\mathcal{T}_{rgb}$ is a YCbCr $\rightarrow$ RGB transformation. Charbonnier loss is commonly used in computer vision, because it's more resilient to outliers compared to MSE. In our experiments, Charbonnier loss didn't improve the final accuracy of the model, but led to faster convergence due to larger gradients.

For optimization, we used AdamW [18] with hyperparameters $\beta_1 = 0.9$, $\beta_2 = 0.999$, $\epsilon = 1 \times 10^{-8}$, weight decay $\lambda = 1 \times 10^{-6}$. The learning rate η is controled by Cosine annealing learning rate scheduler [19] with hyperparameters $\eta_{\text{min}} = 5 \times 10^{-5}$, $\eta_{\text{max}} = 4 \times 10^{-4}$ and $T = 80\ 000$.

Flare Luma and Flare Chroma are trained for a total of 400,000 iterations. The first 20,000 iterations are warmup iterations, during which η is increased linearly from 5×10^{-5} to 4×10^{-4}. Cosine annealing is activated on the 20,000th iteration. Warmup iterations are essential because we don't use normalization layers - without them, gradients exploded and training diverged. We also make use of gradient accumulation for two steps to increase the effective batch size by a factor of two.

We found that training Flare Luma and Flare Chroma simultaneously (end-to-end) yields slightly better final accuracy, so we'll report only results from end-to-end training. We've trained 4 variations of Flare with different numbers of parameters. The number of parameters depends on the number of base channels C and growth factor k. The variants are described in Table 1. For training, we used a workstation with *16-core CPU, 64GB RAM, and NVIDIA RTX A5000 24GB GPU*. The training time varies between 36 h (for the smallest models) and 120 h (for the largest).

Table 1. Flare Variants

	Flare.T	Flare.S	Flare.M	Flare.L
C (Luma)	16	32	48	64
C (Chroma)	16	32	32	64
k (Luma)	2	2	2	2
k (Chroma)	2	1.75	2	1.75
Number of parameters (Luma)	1.0M	4.0M	8.8M	15.7M
Number of parameters (Chroma)	1.0M	2.0M	4.0M	8.0M
Total number of parameters	2.0M	6.0M	12.8M	23.7M

6 Results and Discussion

6.1 Results for Monochrome Images

Many of the first works (and some recent [11]) are developed exclusively for monochrome images, and so we evaluate Flare on monochrome images for comparison. Monochrome images contain only the luminance plane, so the image is run only through Flare Luma (see Sect. 3.1). Figure 6 compares the PSNR of the four variants of Flare Luma for monochrome images from Live1. Even the smallest 1M parameter model can completely remove the block artifacts. Increasing the number of parameters results in improved sharpness, which is most noticeable around the edges. Figure 5 shows reconstructed images with Flare.L model for different q factors. Table 2 compares Flare with previous works for blind q image restoration. Flare achieves the best SSIM (ties with AGARNet).

Table 2. Comparison between models with blind q for monochrome images. Each cell shows PSNR | PSNR-B | SSIM. The best result is in **bold blue**, second best in red.

Dataset	QF	JPEG	QGAC	AGARNet	FBCNN	**Flare**
Classic5	10	27.82|25.21|0.76	29.84|29.43|0.81	30.02|0.00|0.84	**30.12**|**29.80**|0.82	29.83|29.48|**0.83**
	20	30.12|27.50|0.83	31.98|31.37|0.87	32.25|0.00|**0.89**	**32.31**|**31.74**|0.87	32.02|31.46|0.88
	30	31.48|28.94|0.87	33.22|32.42|0.89	33.47|0.00|**0.91**	**33.54**|**32.78**|0.89	33.27|32.45|**0.91**
	40	32.43|29.92|0.89	34.05|33.12|0.91	34.29|0.00|**0.92**	**34.35**|**33.48**|0.91	34.12|33.19|**0.92**
Live1	10	27.77|25.33|0.77	29.51|29.13|0.82	29.64|0.00|**0.84**	**29.75**|**29.40**|0.83	29.55|29.20|**0.84**
	20	30.07|27.57|0.85	31.83|31.25|0.89	32.02|0.00|**0.90**	**32.13**|**31.57**|0.89	31.91|31.37|**0.90**
	30	31.41|28.92|0.89	33.20|32.47|0.91	33.44|0.00|**0.93**	**33.54**|**32.83**|0.92	33.30|32.52|0.92
	40	32.35|29.96|0.90	34.16|33.36|0.93	34.42|0.00|**0.94**	**34.53**|**33.74**|0.93	34.28|33.44|**0.94**

6.2 Results for Color Images

We'll begin our discussion on color images by first motivating the need for a dedicated color image model. A naive approach would be to train a single monochrome model and use it for each color plane. But the reconstruction of color images is additionally complicated by the chroma subsampling. Reconstructing each plane individually may result in slight misalignment between luminance and chrominance, thus chromatic aberration. Another motivation for dedicated color models is the presence of 'false colors' in the output image. When models like ARCNN [27] and DnCNN [28], which are trained only on monochrome images, are adapted for color image reconstruction, they produce colors that weren't present in the original image, and this is particularly noticeable in large gradient areas, such as the sky. Besides that, CbCr planes are quantized with different quantization tables than Y, so a model trained for reconstruction of the Y plane would have suboptimal performance on the reconstruction of the chrominance planes. Flare does not produce false colors, because its

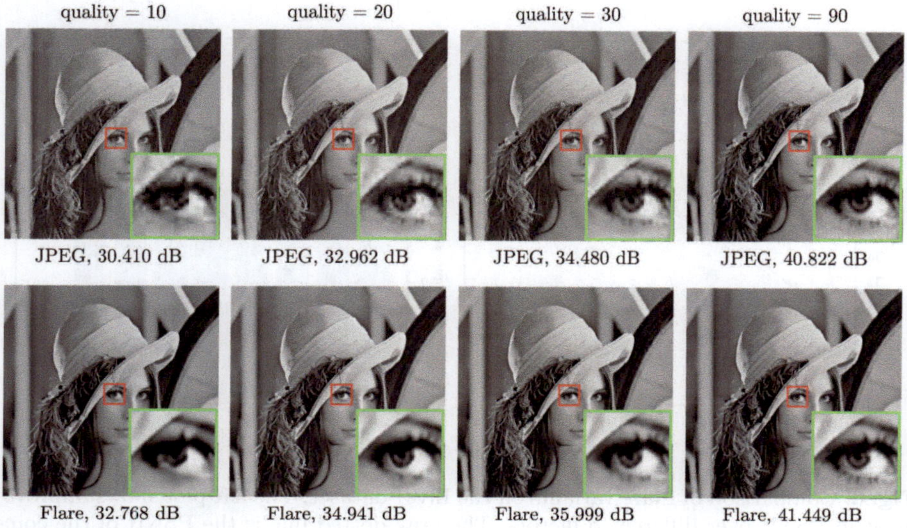

Fig. 5. Reconstruction of 'Lenna' from Classic 5 with Flare Luma. For each image, the PSNR w.r.t. the original is shown. Block artifacts are completely removed without loss of sharpness.

chrominance reconstruction is guided by Y′, which contains more details than CbCr. Figure 7 shows an example reconstruction with Flare for various q for an image from Set14. The PSNR enhancement ranges from $\sim$ 2.5dB to 0.3dB for q $=$ 90. Table 3 compares Flare with other blind methods for block-artifact removal. Flare achieves very close performance to FBCNN [7] while using 3X less parameters (24 million for the largest model vs. 72 million for FBCNN). Figure 8 shows a visual comparison between Flare and FBCNN.

Table 3. Comparison between Flare and other works for color reconstruction. Each cell shows PSNR|PSNR-B|SSIM. Best result is in **bold blue**, second best in red.

<table>
<tr><th>Dataset</th><th>QF</th><th>JPEG</th><th>QGAC</th><th>IDCNN</th><th>FBCNN</th><th>**Flare**</th></tr>
<tr><td>Live1</td><td>10</td><td>25.69|24.20|0.74</td><td>27.62|27.43|0.80</td><td>27.63|**27.63**|**0.82**</td><td>**27.77**|27.51|0.80</td><td>27.73|27.45|**0.82**</td></tr>
<tr><td></td><td>20</td><td>28.06|26.49|0.83</td><td>29.88|29.56|0.87</td><td>30.04|**30.01**|**0.88**</td><td>**30.11**|29.70|0.87</td><td>29.99|29.58|**0.88**</td></tr>
<tr><td></td><td>30</td><td>29.37|27.84|0.86</td><td>31.17|30.77|0.90</td><td>-</td><td>**31.43**|**30.92**|0.90</td><td>31.26|30.71|**0.91**</td></tr>
<tr><td></td><td>40</td><td>30.28|28.84|0.88</td><td>32.05|31.61|0.91</td><td>-</td><td>**32.34**|**31.80**|0.91</td><td>32.14|31.58|**0.92**</td></tr>
<tr><td>BSDS500</td><td>10</td><td>25.84|24.13|0.74</td><td>27.74|27.47|0.80</td><td>27.69|**27.69**|0.81</td><td>**27.85**|27.52|0.80</td><td>**27.85**|27.41|**0.82**</td></tr>
<tr><td></td><td>20</td><td>28.21|26.37|0.83</td><td>30.01|29.53|0.87</td><td>30.07|**30.02**|**0.88**</td><td>**30.14**|29.56|0.87</td><td>textbf30.14|29.40|**0.88**</td></tr>
<tr><td></td><td>30</td><td>29.57|27.72|0.86</td><td>31.33|30.70|0.90</td><td>-</td><td>31.45|**30.72**|0.90</td><td>**31.47**|30.49|**0.91**</td></tr>
<tr><td></td><td>40</td><td>30.52|28.69|0.89</td><td>32.25|31.50|0.92</td><td>-</td><td>32.36|**31.52**|0.91</td><td>**32.39**|31.28|**0.93**</td></tr>
</table>

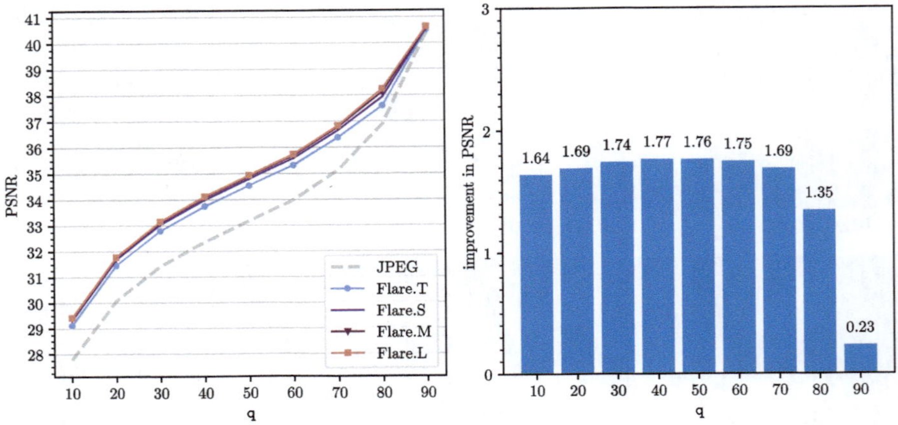

Fig. 6. Comparison of Flare variants on the Live1 dataset. The left plot shows improvement in PSNR over different q factors. The gray dotted line is the PSNR of the compressed JPEG image relative to the original. The right plot shows the improvement in PSNR for the largest model Flare.L with 15.7M parameters. The left plot also shows that additional compression efficiency can be gained when using Flare. For example, an image compressed with $q = 30$ can be restored to an equivalent image that is compressed with $q = 50$. For some images this scheme can achieve $\approx 30\%$ reduction in file size at the expense of more processing power when decoding.

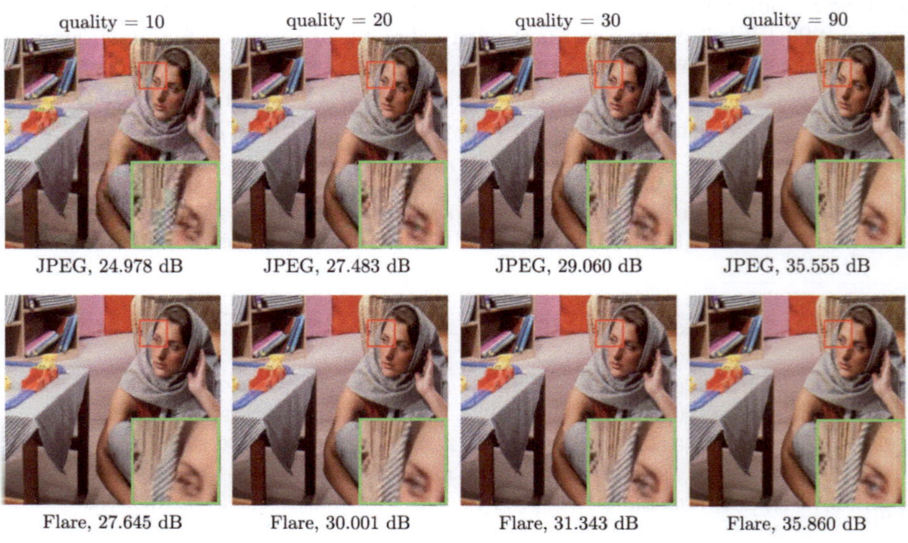

Fig. 7. Reconstruction of 'Barbara' from Set14 with Flare. PSNR is relative to the original image before JPEG compression.

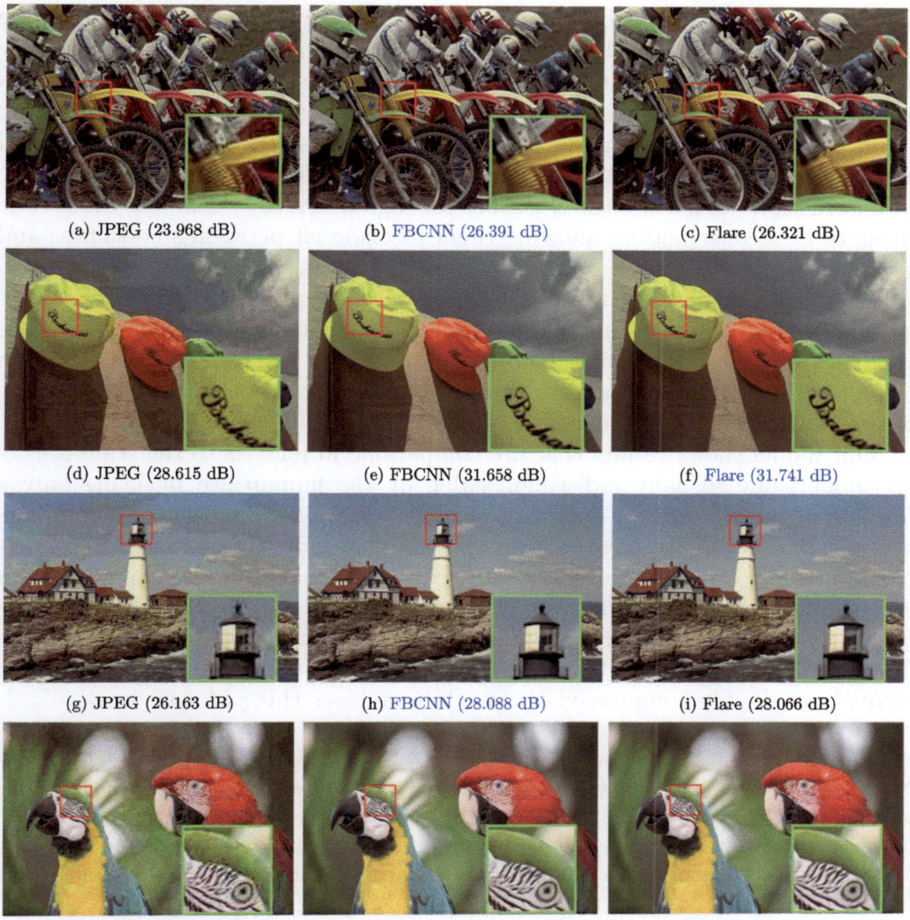

(a) JPEG (23.968 dB) (b) FBCNN (26.391 dB) (c) Flare (26.321 dB)

(d) JPEG (28.615 dB) (e) FBCNN (31.658 dB) (f) Flare (31.741 dB)

(g) JPEG (26.163 dB) (h) FBCNN (28.088 dB) (i) Flare (28.066 dB)

(j) JPEG (29.033 dB) (k) FBCNN (31.932 dB) (l) Flare (32.064 dB)

Fig. 8. Visual comparison of reconstructed images between Flare and FBCNN. The first column shows the compressed JPEG images with $\mathtt{q} = 10$. PSNR is reported under each image and the best PSNR is in blue. The main paper evaluates the performance of reconstruction models in terms of PSNR, PSNR-B and SSIM. These metrics are mathematical functions, that do not align very well to the human vision. The best test is visual appearance and here we see that although FBCNN [7] has higher overall PSNR score on Live1 (see Table 3 in the main text), the reconstructed images look the same. Both models reconstruct smooth gradients very well and preserve details.(Color figure online)

7 Conclusion

In this work, we presented a new neural network for the reduction of block artifacts in JPEG images - **Flare**. Our model is inductively biased on the very algorithm of the JPEG encoder and uses all available information in the JFIF file to reconstruct the image. Flare is a function from the spatial to the frequency domain of the image - it takes the luminance and chrominance image planes as input. It outputs the residuals of the DCT coefficients that were lost during compression. The model can be applied to grayscale and color images compressed with $\mathsf{q} \in [10, 100]$. Moreover, the architecture allows for manual control over the reconstruction - a much-needed feature for digital artists. Users can choose the strength of reconstruction, allowing them to trade off between smoothness and sharpness.

In addition to developing a new model, we've introduced a perceptible-aware q sampling for the generation of input images during training that improves convergence. Instead of sampling compression quality q from uniform distribution U(10, 90), we sample from a distribution which is proportional to the real scaling of the quantization coefficients.

Our model shows results that are comparable in metrics to the state-of-the-art models and visually indistinguishable by the human eye by using only a fraction of the parameters - our largest model, Flare.L 1 has 1/3 of the parameters of FBCNN [7]. We have to note, though, that FBCNN is trained to remove block artifacts even on double-compressed JPEG images, hence the need for a larger number of parameters (Fig. 8). The main difference between FBCNN and Flare is that FBCNN operates entirely in the spatial domain, while Flare is a function from the spatial to the frequency domain. Flare can be retrained to reconstruct doubly compressed images, but we leave this as future work.

Acknowledgments. This work was partially supported by the European Union-NextGenerationEU, through the National Recovery and Resilience Plan of the Republic of Bulgaria [Grant Project No. BG-RRP-2.004-0008]. Views and opinions expressed are, however, those of the authors only and do not necessarily reflect those of the European Union. Neither the European Union nor the granting authority can be held responsible for them.

Declaration on Generative AI. During the preparation of this work, the authors used Grammarly to perform a grammar and spelling check. After using these tool(s)/service(s), the authors reviewed and edited the content as needed and take(s) full responsibility for the publication's content.

1 Visual Comparison

References

1. Agustsson, E., Timofte, R.: NTIRE 2017 challenge on single image super-resolution: dataset and study. In: The IEEE Conference on Computer Vision and Pattern Recognition (CVPR) Workshops (2017)
2. Chen, H., He, X., Qing, L., Xiong, S., Nguyen, T.Q.: DPW-SDNET: dual pixel-wavelet domain deep CNNs for soft decoding of jpeg-compressed images (2018), arXiv:1805.10558

3. Dong, C., Deng, Y., Loy, C.C., Tang, X.: Compression artifacts reduction by a deep convolutional network (2015). arXiv:1504.06993
4. Ehrlich, M., Davis, L., Lim, S.N., Shrivastava, A.: Quantization guided jpeg artifact correction. In: Computer Vision – ECCV 2020: 16th European Conference, Glasgow, UK, August 23–28, 2020, Proceedings, Part VIII, pp. 293–309. Springer-Verlag (2020). https://doi.org/10.1007/978-3-030-58598-3_18
5. Foi, A., Dabov, K., Katkovnik, V., Egiazarian, K.: Shape-adaptive DCT for denoising and image reconstruction. In: Image Processing: Algorithms and Systems, Neural Networks, and Machine Learning, vol. 6064, p. 60640N. International Society for Optics and Photonics, SPIE (2006). https://doi.org/10.1117/12.642839
6. Foi, A., Katkovnik, V., Egiazarian, K.: Pointwise shape-adaptive DCT for high-quality deblocking of compressed color images. In: 2006 14th European Signal Processing Conference, pp. 1–5 (2006)
7. Jiang, J., Zhang, K., Timofte, R.: Towards flexible blind jpeg artifacts removal. In: Proceedings of the IEEE/CVF International Conference on Computer Vision, pp. 4997–5006 (2021)
8. Jung, C., Jiao, L., Qi, H., Sun, T.: Image deblocking via sparse representation. Signal Process. Image Commun. **27**(6), 663–677 (2012). https://doi.org/10.1016/j.image.2012.03.002
9. Kim, Y., Soh, J.W., Cho, N.I.: AGARNet: adaptively gated jpeg compression artifacts removal network for a wide range quality factor. IEEE Access **8**, 20160–20170 (2020). https://doi.org/10.1109/ACCESS.2020.2968944
10. Kwon, Y., Kim, K.I., Tompkin, J., Kim, J.H., Theobalt, C.: Efficient learning of image super-resolution and compression artifact removal with semi-local gaussian processes. IEEE Trans. Pattern Anal. Mach. Intell. **37**(9), 1792–1805 (2015). https://doi.org/10.1109/TPAMI.2015.2389797
11. Liang, J., Cao, J., Sun, G., Zhang, K., Van Gool, L., Timofte, R.: SwinIR: image restoration using swin transformer. In: 2021 IEEE/CVF International Conference on Computer Vision Workshops (ICCVW), pp. 1833–1844 (2021). https://doi.org/10.1109/ICCVW54120.2021.00210
12. Liew, A.C., Yan, H.: Blocking artifacts suppression in block-coded images using overcomplete wavelet representation. IEEE Trans. Circuits Syst. Video Technol. **14**(4), 450–461 (2004). https://doi.org/10.1109/TCSVT.2004.825555
13. Lim, B., Son, S., Kim, H., Nah, S., Lee, K.M.: Enhanced deep residual networks for single image super-resolution. In: The IEEE Conference on Computer Vision and Pattern Recognition (CVPR) Workshops (2017)
14. List, P., Joch, A., Lainema, J., Bjontegaard, G., Karczewicz, M.: Adaptive deblocking filter. IEEE Trans. Circuits Syst. Video Technol. **13**(7), 614–619 (2003). https://doi.org/10.1109/TCSVT.2003.815175
15. Liu, P., Zhang, H., Zhang, K., Lin, L., Zuo, W.: Multi-level wavelet-CNN for image restoration (2018). arXiv:1805.07071
16. Liu, X., Wu, X., Zhou, J., Zhao, D.: Data-driven sparsity-based restoration of jpeg-compressed images in dual transform-pixel domain. In: 2015 IEEE Conference on Computer Vision and Pattern Recognition (CVPR), pp. 5171–5178 (2015). https://doi.org/10.1109/CVPR.2015.7299153
17. Liu, Z., et al.: Swin Transformer: hierarchical vision transformer using shifted windows. In: 2021 IEEE/CVF International Conference on Computer Vision (ICCV), pp. 9992–10002 (2021). https://doi.org/10.1109/ICCV48922.2021.00986
18. Loshchilov, I., Hutter, F.: Fixing weight decay regularization in ADAM (2017). http://arxiv.org/abs/1711.05101

19. Loshchilov, I., Hutter, F.: SGDR: stochastic gradient descent with warm restarts (2017). arXiv:1608.03983
20. Perez, E., Strub, F., de Vries, H., Dumoulin, V., Courville, A.: FiLM: visual reasoning with a general conditioning layer (2017). arXiv:1709.07871
21. Ronneberger, O., Fischer, P., Brox, T.: U-Net: convolutional networks for biomedical image segmentation (2015). arXiv:1505.04597
22. Rothe, R., Timofte, R., Van Gool, L.: Efficient regression priors for reducing image compression artifacts. In: 2015 IEEE International Conference on Image Processing (ICIP), pp. 1543–1547 (2015). https://doi.org/10.1109/ICIP.2015.7351059
23. Sun, D., Cham, W.K.: Postprocessing of low bit-rate block DCT coded images based on a fields of experts prior. IEEE Trans. Image Process. **16**(11), 2743–2751 (2007). https://doi.org/10.1109/TIP.2007.904969
24. Tolstaya, E., Rychagov, M., Kim, S., Choi, D.: Removal of blocking and ringing artifacts in JPEG-coded images. In: vol. 7537, p. 75370 (2010). https://doi.org/10.1117/12.838447
25. Wang, Z., Liu, D., Chang, S., Ling, Q., Yang, Y., Huang, T.S.: D3: deep dual-domain based fast restoration of JPEG-compressed images. In: 2016 IEEE Conference on Computer Vision and Pattern Recognition (CVPR), pp. 2764–2772 (2016). https://doi.org/10.1109/CVPR.2016.302
26. Yang, Y., Galatsanos, N., Katsaggelos, A.: Projection-based spatially adaptive reconstruction of block-transform compressed images. IEEE Trans. Image Process. **4**(7), 896–908 (1995). https://doi.org/10.1109/83.392332
27. Yu, K., Dong, C., Loy, C.C., Tang, X.: Deep convolution networks for compression artifacts reduction (2016). arXiv:1608.02778
28. Zhang, K., Zuo, W., Chen, Y., Meng, D., Zhang, L.: Beyond a gaussian denoiser: residual learning of deep CNN for image denoising. IEEE Trans. Image Process. **26**(7), 3142–3155 (2017). https://doi.org/10.1109/tip.2017.2662206
29. Zhang, X., Yang, W., Hu, Y., Liu, J.: DMCNN: dual-domain multi-scale convolutional neural network for compression artifacts removal. In: 2018 25th IEEE International Conference on Image Processing (ICIP), pp. 390–394 (2018). https://doi.org/10.1109/ICIP.2018.8451694
30. Zheng, B., Chen, Y., Tian, X., Zhou, F., Liu, X.: Implicit dual-domain convolutional network for robust color image compression artifact reduction. IEEE Trans. Circuits Syst. Video Technol. **30**(11), 3982–3994 (2020). https://doi.org/10.1109/tcsvt.2019.2931045
31. Zini, S., Bianco, S., Schettini, R.: Deep residual autoencoder for blind universal jpeg restoration. IEEE Access **8**, 63283–63294 (2020). https://doi.org/10.1109/access.2020.2984387

Dynamic User Segmentation via Machine Learning for Personalized Mobile App Engagement

Varlamova Mariia(✉) and Okolo-Kulak Polina

Graduate School of Management, St. Petersburg University, St. Petersburg 199004, Russia
vmp2808@yandex.ru

Abstract. The rapid growth of mobile applications and rising competition for user attention necessitates precise segmentation. Traditional methods, relying on demographic or basic behavioral features, offer limited predictive power and cannot anticipate individual responses to personalized offers.

This study proposes and evaluates a machine learning framework for real-time, behavior-driven user segmentation, designed to reduce push and server costs in mobile apps. The approach combines temporal quantitative features engineering with unsupervised pattern discovery in user session sequences, enabling segments to adapt weekly to evolving user trajectories.

A probabilistic, data-driven method dynamically assigns users to segments based on their predicted likelihood to perform target actions over 7, 14, or 30-day horizons. An interpretable ensemble machine learning model ensures both high predictive accuracy and transparency, while segmentation thresholds are dynamically adjusted to align with diverse business objectives and interaction types.

Experimental evaluation on the company's mobile app usage datasets demonstrates that the proposed framework effectively prioritizes users based on their expected value. The model achieves a ROC-AUC score exceeding 0.80, enabling more efficient resource allocation and strategic messaging.

The main contribution is a scalable approach to user segmentation for mobile offerings, integrating predictive modeling and probabilistic segmentation into a scalable, real-time framework for adaptive personalization, improved user engagement, and monetization, providing a practical tool for data-driven decision-making in competitive digital environments.

Keywords: dynamic user segmentation · predictive modeling · ensemble learning · digital behavior analysis · machine learning-based personalization

1 Introduction

Modern mobile applications face growing competition for user attention, emphasizing the need for advanced personalization and effective engagement strategies. However, existing segmentation approaches, primarily based on demographic data or aggregated

I. K. Koychev (Ed.): ERIS 2025, PINNS 4, pp. 90–105, 2026.
https://doi.org/10.1007/978-3-032-30153-6_6

behavioral metrics, provide only static insights. These methods lack the ability to model temporal dynamics, contextual factors, and fine-grained behavioral variations, leading to limited predictive power and suboptimal targeting. Current literature offers few scalable frameworks capable of dynamically segmenting users based on evolving behavior patterns and predicting short-term engagement or transactional actions. This gap highlights the need for machine learning–driven segmentation models that incorporate temporal behavioral features and probabilistic predictions to enable proactive, data-driven engagement strategies.

The need for more accurate and adaptive segmentation methods is driven by the exponential growth in the volume of digital data. For instance, the global data volume reached 149 zettabytes in 2024 and is projected to grow to 181 zettabytes by 2025, equivalent to generating 2.5 quintillion bytes of data every day [1]. This volume and velocity of data require new approaches to analysis and segmentation capable of processing complex and rapidly changing behavioral patterns.

The aim of this research is to develop a machine learning–based framework for dynamic user segmentation that enhances personalization and engagement in mobile applications. To achieve this goal, the following research objectives were set:

1. To analyze existing approaches to user segmentation in digital environments and identify their limitations.
2. To design a probabilistic segmentation framework that incorporates both quantitative activity logs and qualitative behavioral aspects.
3. To implement and evaluate ensemble machine learning models for predicting user actions and dynamically forming user segments.
4. To validate the proposed framework through experiments on real-world mobile application datasets and assess its practical applicability.

The theoretical and methodological foundation of this thesis is based on fundamental works and theories in the fields of behavioral economics, machine learning algorithms for user segmentation and data-driven personalization approaches in mobile apps. Specifically, the works of the following scholars and researchers were utilized and studied: machine learning (Leo Breiman), data-driven personalization (Rohit Sharma). Also, scientific articles and electronic resources with research papers of various data scientists and other practitioners in the sphere of machine learning.

Thus, the following propositions, which will be defended, were formulated in this article:

1. User behavior in the digital environment of mobile applications is not random, as it exhibits stable patterns that correlate with decision-making outcomes. Consequently, behavioral patterns extracted from historical activity log data within the mobile application contain statistically significant predictors that can be used to accurately forecast the likelihood of a target action being performed.
2. Dynamic segmentation, based on probabilistic modeling and ensemble machine learning, outperforms static segmentation methods by enabling real-time adaptation to user behavior and delivering higher predictive accuracy for engagement outcomes.

The practical significance of the research lies in the fact that the proposed model and personalization algorithms can be integrated into mobile applications to enhance user engagement and retention. The developed methodology can be adapted to different domains such as fintech, govtech, and service automation, where user interaction patterns are dynamic and data-intensive. Furthermore, the research findings can serve as a foundation for further scientific and applied studies related to personalizing user experiences, optimizing resource allocation, and implementing machine learning in digital services. At a broader level, the approach contributes to the advancement of data-driven personalization practices, offering insights into how predictive modeling can support scalable and ethically responsible engagement strategies in competitive digital markets.

2 Literature Review

Early research on user segmentation with known class labels employed classification algorithms. Logistic regression provided interpretable results but had a limited capacity to model non-linear dependencies [2]. Decision trees and ensemble methods, including random forests, enhanced model robustness by combining multiple predictors; however, with limited data, they were prone to overfitting [3]. Support vector machines demonstrated high accuracy with non-linear features but required substantial computational resources and careful hyperparameter tuning [4]. A primary limitation of these methods is their dependence on labeled data, whose collection and annotation are labor-intensive.

Clustering methods enabled the identification of latent groups without labels. The K-Means algorithm effectively segmented customers based on transactional activity but required a pre-defined number of clusters [5]. Hierarchical clustering and DBSCAN were applied to analyze payment behavior and detect anomalies; however, they are sensitive to parameter selection and feature quality [6].

With the adoption of deep learning in segmentation, neural network-based methods began to be used for the automatic discovery of complex patterns in large datasets. Autoencoders reduced dimensionality and extracted latent features, thereby improving clustering quality. Recurrent Neural Networks (RNNs) analyzed behavioral sequences, including the temporal dynamics of transactions and in-app activity [7]. Despite the progress achieved, these methods face challenges related to scalability, high data variability, and the need to adapt to rapidly changing user behavioral patterns.

3 Data Sources

This study outlines a methodology for user segmentation within the RAY digital ecosystem, a platform providing integrated services for motorists, including its core "Traffic Fines" application for violation payments, e-OSAGO, vehicle insurance, and roadside assistance.

The analysis utilizes user interaction data sourced from the "Traffic Fines" mobile application. Raw event logs are captured and stored in a Yandex AppMetrica cloud storage bucket (S3-compatible API) owned by RAY. For processing, this data was transferred to the SIMBA big data platform (S3 storage) at GSOM, St. Petersburg State University. Data ingestion was performed via API requests using an access token, with event logs structured as daily batches. The dataset comprises detailed user session logs, including application events, push notification interactions, and device metadata (e.g., manufacturer, OS, connection type). Each event is defined by a timestamp, a unique event name, and a user profile identifier.

Given the scale and velocity of data, distributed preprocessing was implemented in PySpark. For feature engineering, the most informative attributes were identified as:

- profile_id (user identifier),
- event_datetime (timestamp),
- event_name (categorical descriptions of user actions, e.g., Payment accepted, Completed payment, Card entry, Push notification, Viewing fine details).

This structure enabled the construction of behavioral sequences and extraction of predictors relevant to payment probability modeling.

4 Methodology

4.1 Research Objective

The research objective was to estimate the probability of a user completing a fine payment in the application within an annual observation period. Target events were defined as "Payment accepted" or "Payment completed" and encoded as a binary outcome (0 – other events, 1 – payment).

4.2 Event Vectorization

For event vectorization, the Term Frequency–Inverse Document Frequency (TF-IDF) method [8] was employed due to its domain-specific advantages. Its selection was motivated by three factors:

1. the keyword-driven nature of payment events, effectively captured by TF-IDF through weighted term representation;
2. the brevity of mobile application events, where TF-IDF often outperforms contextual embeddings such as Word2Vec or BERT;
3. the capacity of TF-IDF to amplify rare but highly informative behavioral signals.

For comparison, alternative embedding methods (Word2Vec, BERT) were also implemented; however, their application did not provide measurable performance benefits given the limited contextual richness of the dataset and their computational cost.

4.3 Class Imbalance

The dataset exhibited a strong imbalance between payment and non-payment events. To address this, random undersampling was applied, balancing the training distribution and reducing bias towards the majority class [9]. This increased sensitivity to subtle behavioral patterns (e.g., repeated access to payment pages, push notification interactions). The Synthetic Minority Oversampling Technique (SMOTE) was additionally tested [10]. However, it failed to yield significant improvements, likely due to the sequential and context-dependent nature of user interactions that synthetic instances could not adequately replicate [11].

4.4 Feature Engineering

Feature engineering was conducted to capture user behavioral dynamics within the application. User actions were aggregated across multiple temporal windows preceding the target event, ranging from short-term (10 min–1 h, 1–24 h, 1–3 days, 3–7 days) to long-term periods (up to 6 months). This design enabled the model to account for both immediate and cumulative interaction patterns.

Special attention during research is paid to such an event as a push notification, since it is this event that often encourages users to pay, especially the notification of the end of the discount on the fine. Based on this event, separate features are created that reflect the total number of notifications in each time window, if such events have been observed for each specific user.

Several experimental configurations were tested by varying combinations of time windows. While shorter windows (e.g., up to 7 days) captured highly localized behaviors, they were deemed less practical from a business perspective, as they provide insufficient lead time for proactive interventions (e.g., targeted notifications or marketing campaigns to push the user to pay).

Also, the following model features were tested: binary factors indicating the presence or absence of push notifications, as well as similar binary factors for different types of push messages (e.g., a fine payment reminder, a discount expiration reminder). Additionally, to evaluate the relevance of including additional features in the model, statistical tests (Two-sample Z-test for proportions) for significance were conducted on parameters such as mobile operator, day of the week of the payment, and time of day of the payment. Unfortunately, these features appeared to be irrelevant thus they have not been included into the final model.

The features combination for the final model is shown on Fig. 1 below.

payment_event_flag	event_name	lag_3days_to_7days	lag_7days_to_15days	lag_15days_to_30days	lag_1mth_to_2mth	push_count_3days_to_7days	push_count_7days_to_15days	push_count_15days_to_30days	push_count_1mth_to_2mth
1	Мои штрафы/ Оплата/ Завершили оплату	[Мои штрафы/ Оплата/Завершили оплату, Мои штраф...	[Проверка/Список штрафов/ГИС ГМП проверена, Мо...	[Мои штрафы/ Документы/Открытие документа, Мои ...	[Мои штрафы/ Оплата/Платёж принят, Мои штрафы/О...	2	1	3	6
1	Мои штрафы/ Оплата/ Завершили оплату	[Мои штрафы/ Оплата/Завершили оплату, Мои штраф...	[Проверка/Список штрафов/ГИС ГМП проверена, Мо...	[Мои штрафы/Оплата/ Платёж принят, Мои штрафы/О...	[Мои штрафы/ Оплата/Платёж принят, Мои штрафы/О...	2	1	3	6
1	Мои штрафы/ Оплата/ Платёж принят	[Мои штрафы/ Оплата/Платёж принят, ОСАГО/Дата п...	[Мои штрафы/ Документы/Есть документ, Push/ Диал...	[Запуск, Мои штрафы/ Проверка/ГИС ГМП проверена...	[Мои штрафы/ Оплата/Завершили оплату, Мои штраф...	2	1	3	6
1	Мои штрафы/ Оплата/ Завершили оплату	[Мои штрафы/ Оплата/Завершили оплату, Мои штраф...	[Запуск, Мои штрафы/Оплата/ Платёж принят, Мои ...	[Мои штрафы/Оплата/ Платёж принят, Мои штрафы/О...	[Проверка/Список штрафов, Штраф, Мои штрафы/Оп...	2	1	3	6
1	Мои штрафы/ Оплата/ Завершили оплату	[Мои штрафы/ Оплата/Завершили оплату, Мои штраф...	[Распознавание ВУ/ Открыта подсказка с камеры, ...	[Мои штрафы/Оплата/ Платёж принят, Мои штрафы/О...	[Сторис/Пост/ Открыт, Мои штрафы/Оплата/ Платёж ...	2	1	3	6

Fig. 1. Features in the final model.

4.5 Model Selection and Training

The training period has been chosen from 01-01-2024 to 31-12-2024. The tested algorithms are as follows: Logistic regression, Random forest, SVM, Gradient Boosting Trees, Light Gradient Boosting Model, CatBoostClassifier, XGBClassifier.

To improve the performance of the models, hyperparameters have been configured using a classic method such as GridSearch with cross-validation, and a more advanced one such as Hyperopt (python library). The GridSearch approach is quite slow, since it is necessary to search through all combinations of all parameters, and the search will continue even with obviously unsuccessful combinations. Hyperopt is based on Bayesian optimization. The hyperparameter values in the current iteration are selected based on the results of the previous step. The main idea of the algorithm is as follows: at each iteration of the selection, there is a compromise between the study of regions with the most successful combinations of hyperparameters and the study of regions with high uncertainty (where even more successful combinations can be found). This makes it possible in many cases to find the best values of the model parameters in less time.

For training and evaluating models, a standard approach has been chosen, which consists in dividing the data into 80% for the training dataset and 20% for the test dataset. The number of events in the dataset of the two classes for evaluation of subsequent models is shown in Table 1:

Table 1. Classes and number of events.

Classes	Number of events
Class 0 (Non-Payment)	292 100
Class 1 (Payment)	204 948

4.6 Model Evaluation

The following metrics have been selected as key indicators for evaluating the model's performance: ROC-AUC, PR-Curve, Precision, Recall, and F1-Score [12]. These metrics were chosen to address the challenges of imbalanced classification. ROC-AUC provides a threshold-independent measure of overall class separability, while PR-based metrics and F1-Score focus on the model's performance on the minority class (successful payments), preventing bias toward the majority class. Precision and Recall are defined as:

$$\text{Precision} = \frac{TP}{(TP + FP)} \tag{1}$$

$$\text{Recall} = \frac{TP}{(TP + FN)} \tag{2}$$

The F1-score, representing the harmonic mean of Precision and Recall, is given by:

$$\text{F1} - \text{score} = \frac{2 * Precision * Recall}{(Precision + Recall)} \tag{3}$$

The ROC-AUC measures the model's ability to discriminate between positive and negative classes across all possible classification thresholds. The ROC curve plots the

True Positive Rate (TPR), which is equivalent to Recall, against the False Positive Rate (FPR), defined as:

$$\mathrm{FPR} = \frac{FP}{(FP + TN)} \tag{4}$$

ROC-AUC is defined as the area under the ROC curve and ranges from 0.5, indicating random classification, to 1.0, representing perfect class separation. Due to its threshold-independent nature, ROC-AUC enables robust comparison of models when the optimal decision threshold is unknown.

The PR-Curve illustrates the relationship between Precision and Recall across different classification thresholds, focusing specifically on the positive class. The area under the PR-Curve (PR-AUC) is used as a summary metric of this curve. In contrast to ROC-AUC, PR-AUC is sensitive to class imbalance and provides a more informative assessment when the positive class represents a small fraction of the dataset. A perfect classifier achieves a PR-AUC of 1, whereas a random classifier's PR-AUC is proportional to the prevalence of the positive class.

Here, TP (true positive) indicates the number of correctly predicted positive events, i.e. payment events, FP (false positive) refers to negative events (non-payments) that have been incorrectly predicted as positive, FN (false negative) indicates positive events that have been incorrectly predicted as negative [13].

These indicators provide a comprehensive assessment of the model's ability to distinguish classes, especially in conditions of data imbalance. From a business perspective, Recall was prioritized, as it is critical to identify as many potential payers as possible. Maximizing recall reduces the risk of missed revenue from undetected paying users and increases the effectiveness of marketing interventions, such as targeted push notifications.

User segmentation was performed by generating predictions on a hold-out validation sample using the best-performing model. For operational forecasting, the model requires periodic retraining on the most recent data to account for evolving user behavior and maintain predictive accuracy. This necessity for retraining underscores the dynamic nature of the segmentation, as it reflects current behavioral patterns.

5 Results and Analysis

5.1 Comparative Analysis of Machine Learning Models in the PySpark Environment

All experimental computations were carried out in JupyterLab on a workstation with an 8 GB CPU-based configuration. Despite relatively limited hardware resources, the workflow was designed to efficiently process large-scale event logs from the mobile application. The analysis pipeline consisted of two stages: initial preprocessing with PySpark to handle large datasets, and subsequent model development and experimentation in JupyterLab using Python-based machine learning libraries.

Several machine learning algorithms were implemented in PySpark and their results were compared to predict the likelihood of user payment within a mobile application. As a baseline, Logistic Regression was used due to its simplicity and interpretability [14]. LightGBM consistently demonstrated superior performance, offering the best trade-off between predictive accuracy, scalability, and computational efficiency, making it particularly suitable for user payment probability in a mobile app [15].

The comparative performance of all models is summarized in Table 2. Logistic Regression showed poor predictive power for the minority class (paying users), achieving a recall of only 0.15, which highlights the limitations of linear models for highly imbalanced behavioral data. In contrast, tree-based ensemble methods such as Random Forest, Gradient Boosting Trees, and especially LightGBM achieved substantially higher recall and balanced F1-scores [16]. Notably, LightGBM outperformed other classifiers in terms of ROC-AUC (0.8365), indicating stronger overall discriminative ability.

Table 2. Model performance comparison.

Model	Class	Recall	F1-Score	ROC-AUC
LR	0	0.852	0.7193	0.512
	1	0.1512	0.324	
SVM	0	0.7949	0.7946	0.8168
	1	0.7361	0.7365	
RFC	0	0.8026	0.7962	0.8177
	1	0.7265	0.7341	
GBT	0	0.7801	0.8004	0.8242
	1	0.7834	0.7587	
LightGBM	**0**	**0.7736**	**0.8001**	**0.8365**
	1	**0.7947**	**0.7624**	

These results suggest that gradient boosting methods are more effective at capturing non-linear dependencies and interactions between user features, which are critical for accurately estimating the probability of fine payment. In particular, the higher recall values obtained with boosting-based approaches are essential for reducing false negatives, ensuring that users with a high likelihood of completing the target action are not overlooked. This makes the model not only more precise in segmentation but also more reliable for decision-making in real-world application scenarios.

5.2 Additional Gradient Boosting Experiments in Jupyter Notebook

To complement the PySpark-based experiments, additional tests were conducted in Jupyter Notebook. The motivation for this stage was the inability to train certain advanced gradient boosting models, such as CatBoost and XGBoost, directly within the PySpark environment. Therefore, a separate workflow was designed to enable evaluation of these algorithms on a smaller but representative dataset.

A one-month observation period (March 2025) was selected, yielding 22,416 records that included uncoded lag sequences of user actions along with counts of push notifications. After cleaning and lemmatization of lag features, text vectorization was performed using TF-IDF, and numerical features were standardized. This dataset was then used to train and compare multiple classifiers, including Logistic Regression, Decision Tree, CatBoost, XGBoost, and LightGBM [17].

To address the problem of severe class imbalance, both built-in balancing techniques and external resampling strategies were applied. Experiments with RandomUnderSampler and SMOTE demonstrated that synthetic oversampling significantly improved the detection of positive outcomes [18]. For example, the LGBMClassifier combined with SMOTE achieved the highest performance (F1-score = 0.5898, ROC-AUC = 0.7268), while morphological preprocessing with TF-IDF further improved the F1-score to 0.6011. These results, summarized in Table 3, highlight the sensitivity of model quality to text preprocessing choices.

Table 3. Results of additional model testing in Jupyter Notebook.

Model	ROC-AUC	F1-Score	Time
WordNetLemmatizer, TF-IDF			
LogisticRegression	0.7049	0.5560	1 min 34 s
DecisionTreeClassifier	0.7019	0.5561	12 min 22 s
CatBoostClassifier	0.7286	0.5302	23 min 47 s
XGBClassifier	0.7230	0.4988	19 min 59 s
LGBMClassifier	0.7286	0.5819	19 min 1 s
LogisticRegression + RandomUnderSampler	0.7025	0.5453	7 min 48 s
LogisticRegression + SMOTE	0.7063	0.5598	28 min 15 s
LGBMClassifier + SMOTE	**0.7268**	**0.5898**	**31 min 18 s**
MorphAnalyzer, TF-IDF			
LGBMClassifier + SMOTE	0.7112	0.6011	11 min 11 s
Processing text columns with BERT			
LogisticRegression	0.6998	0.5785	2 min 10 s
LGBMClassifier + SMOTE	0.7167	0.5680	1 h 36 min 4 s

Although models trained in Jupyter generally underperformed compared to the Spark-based pipeline (where ROC-AUC exceeded 0.8), these experiments provided several key insights. First, boosting models such as CatBoost and XGBoost proved competitive when combined with proper resampling, confirming their potential for the task. Second, the experiments revealed the importance of preprocessing methods: using morphological analysis yielded a significant gain over simple lemmatization. Finally, attempts to integrate modern embedding techniques like BERT demonstrated promise but were constrained by hardware limitations, as resource requirements exceeded the capabilities of the experimental setup.

To explicitly assess the trade-off between predictive performance and computational efficiency under hardware constraints, a runtime benchmark of classical and embedding-based approaches was conducted. As shown in Table 3, models relying on traditional text representations such as TF-IDF combined with linear or gradient boosting classifiers achieved competitive F1-scores while requiring significantly less training time compared to transformer-based embeddings.

In particular, Logistic Regression with TF-IDF completed training in under two minutes, whereas the use of BERT embeddings increased runtime by more than an order of magnitude, reaching over 1.5 h for the LGBMClassifier, with only marginal gains in predictive performance. These results demonstrate that, under limited computational resources, simpler text representations offer a favorable balance between runtime efficiency and model quality, making them more suitable for large-scale or time-sensitive applications.

5.3 Final Testing of the Best Model and Feature Importance Analysis

Based on the results of previous model development stages, the best-performing model was identified as the LightGBM created in Pyspark. This model was subsequently tested on a holdout dataset, where it correctly identified 7,685 users who paid fines (TP) and 9,563 users who did not pay (TN), while incorrectly classifying 2,804 non-paying users as payers (FP) and failing to detect 1,985 actual payers (FN). These results highlight the model's strong ability to capture the majority of paying users, which is crucial for minimizing missed target actions, while the impact of false positives is comparatively less critical.

A key advantage of gradient boosting methods, including LightGBM, lies in their capacity to provide interpretable rankings of feature importance. Analysis of the top predictors revealed that the most informative signals originate from user activity within the 7–30 day lag windows preceding a payment. Features associated with the 15–30 day window exhibited the highest contribution, suggesting that users typically require a latency period before completing fine payments. Events such as prior payment completions or reviewing historical receipts consistently ranked among the strongest predictors, indicating that users with established transactional history form a segment with elevated payment propensity.

These findings not only validate the predictive performance of the proposed LightGBM-based framework but also establish a robust foundation for dynamic user segmentation. The interpretability of temporal features—especially the relevance of behavioral patterns emerging in the 7–30 day pre-payment window—enables the translation of probabilistic model outputs into actionable business strategies. By leveraging these insights, users can be dynamically classified into payment propensity segments, supporting proactive engagement campaigns and optimized resource allocation. Thus, the model functions as both a high-performance predictive tool and a decision-support system for strategic targeting in mobile application environments.

5.4 Threshold Selection and User Segmentation

Following model evaluation, users were segmented based on the predicted probability of same-day fine payment generated by the best-performing LightGBM model on the verification sample (January–March 2025).

Given the pronounced class imbalance, the optimization strategy prioritized Recall to minimize missed payments. Using the default 50% threshold produced a low Recall of 51% for actual payers, whereas lowering the threshold to 40% substantially increased the detection of positive cases, which is essential for maximizing revenue capture. This adjustment balances the trade-off between identifying potential payers and limiting false positives, thereby enabling more efficient targeting of push notifications compared to indiscriminate campaigns.

The effectiveness of the chosen threshold was validated using the Lift metric. Users ranked by predicted probability in the top 30% (corresponding to the 40% threshold) exhibited a conversion rate of 70.05%, yielding a Lift of 2 relative to random selection. This demonstrates the model's strong discriminative power and supports threshold-based segmentation for operational decision-making.

Based on the 40% threshold and the distribution of predicted probabilities for actual payers (mean ~ 70%), users were classified into three segments:

- High-probability payers (70–100%): Core, loyal users with strong intent to pay.
- Medium-probability (potential) payers (40–69%): Users with moderate intent, representing a key target for conversion through tailored engagement.
- Low-probability payers (<40%): Users with low likelihood of immediate payment based on current behavioral signals.

This probabilistic segmentation framework enables company to dynamically allocate engagement resources, focusing on high-value users while applying targeted interventions to the medium-probability segment to increase conversion rates.

6 Discussion

6.1 Performance Evaluation of the Proposed Framework

The proposed framework effectively achieves the research objective of developing an adaptive and data-driven user segmentation method for mobile applications. The LightGBM model, trained on time-windowed behavioral features and push notification data, outperformed baseline models such as Logistic Regression and performed competitively against other ensemble classifiers, achieving a ROC-AUC above 0.83 and high Recall for the minority (payer) class. Feature analysis confirmed that behavioral indicators—including frequent visits to the payment screen, responsiveness to notifications, and interaction frequency—serve as statistically significant predictors of fine payment likelihood, supporting Proposition 1.

The segmentation strategy operationalizes predicted probabilities through empirically optimized thresholds (e.g., 40% and 70%), enabling precise targeting of different user cohorts. Validation using the Lift metric demonstrates that the framework delivers substantially higher returns compared to random targeting. Moreover, the company's segmentation approach, based primarily on statistical behavioral patterns from customer support interactions, lacked predictive capabilities regarding users' future engagement or fine payment behavior. By contrast, the proposed probabilistic segmentation model accurately identifies users' short-term payment likelihood, providing actionable insights for marketing campaigns. This supports Proposition 2: dynamic segmentation based on probabilistic modeling and ensemble learning outperforms static approaches, offering adaptability and improved predictive accuracy.

The model also provides actionable business value. High-probability users (≥70%) can be targeted with retention campaigns, medium-probability users (40–69%) with personalized nudges, and low-probability users (<40%) deprioritized to optimize resource allocation. The model's design supports a 7-day forecast horizon, aligned with the training methodology, which can be adjusted to meet strategic needs by retraining with time-shifted features. This flexibility enables proactive campaigns rather than reactive engagement, improving marketing efficiency [19].

6.2 Limitations and Future Research Directions

Despite promising results, several limitations warrant consideration. First, the framework relies primarily on dynamic behavioral features from event logs and notifications, excluding static attributes such as demographics or socio-economic indicators. Integrating quasi-static variables (e.g., cumulative payment history) or developing a hybrid model combining dynamic and static features could enhance robustness and improve long-term predictive accuracy.

Second, model performance depends on frequent retraining to capture evolving user behavior. Future work should focus on developing automated retraining pipelines, drift detection mechanisms, and performance monitoring systems to ensure stability in production.

Third, segmentation thresholds are empirically defined and involve trade-offs between recall and precision. Investigating cost-sensitive threshold optimization or reinforcement learning-based strategies may improve decision-making under resource constraints.

Finally, while LightGBM provides feature importance rankings, it lacks causal interpretability. Establishing causal links between user actions (e.g., viewing payment reminders) and payments would enable more targeted interventions. Future research could incorporate causal inference techniques or counterfactual modeling to enhance decision support.

6.3 Impact of Dynamic Probabilistic Segmentation on Personalization and User Engagement

The proposed machine learning–based framework for dynamic user segmentation enhances personalization and user engagement in mobile applications by translating probabilistic predictions into real-time, actionable insights. Unlike traditional segmentation approaches that rely on fixed user attributes, the framework continuously updates user segments based on recent behavioral signals and predicted payment propensity. This dynamic nature enables personalization strategies to adapt in near real time to changes in user intent, attention, and readiness to act.

From a personalization perspective, the framework operationalizes predicted payment probabilities as a central decision variable for engagement design. Rather than treating all users uniformly, the model estimates each user's likelihood of completing a payment within a short-term forecast horizon and maps these estimates to actionable segments (high, medium, and low propensity). This segmentation enables the application to tailor the tone, timing, and intensity of user communications.

Specifically, users with a high predicted payment probability can be engaged through concise reminder notifications or confirmation-style messages that require minimal effort, whereas users in the medium-probability segment can receive informational or motivational messages aimed at reducing potential barriers to payment, such as uncertainty regarding deadlines or payment procedures. Users with low predicted payment probability may be temporarily excluded from proactive push notification campaigns, thereby limiting unnecessary communication and mitigating notification fatigue, which is critical for maintaining long-term user engagement.

In terms of user engagement, the dynamic nature of the segmentation framework allows interaction strategies to adapt to temporal fluctuations in user behavior. The empirical results demonstrate that short-term behavioral patterns—particularly those emerging within the 7–30 day window preceding payment—carry substantial predictive power. By incorporating these temporally sensitive signals, the framework enables engagement interventions to be aligned with periods of heightened user receptivity, thereby increasing the likelihood of meaningful interaction without increasing overall communication volume.

Importantly, the framework supports continuous re-segmentation as user behavior evolves. Users can transition between segments in response to new actions or changes in interaction frequency, allowing engagement strategies to adjust automatically without manual intervention. This adaptability is particularly critical in mobile application contexts, where user intent is volatile and static segmentation rapidly loses relevance.

Overall, the proposed framework establishes a direct link between predictive modeling and engagement execution. By embedding probabilistic user segmentation into the personalization pipeline, the system translates behavioral insights into scalable, data-driven engagement strategies that improve both operational efficiency and user experience.

7 Conclusion

This study presents the development and validation of a dynamic, machine learning–based framework for user segmentation in mobile applications, addressing the need for adaptive personalization in competitive digital ecosystems. Traditional segmentation approaches, often reliant on demographic attributes or aggregated behavioral metrics, fail to capture the temporal and contextual complexity of mobile app interactions. The proposed framework leverages detailed event logs and push notification data to accurately forecast user actions, specifically the likelihood of fine payments.

The proposed methodology for user segmentation consists of the following main stages (Fig. 2):

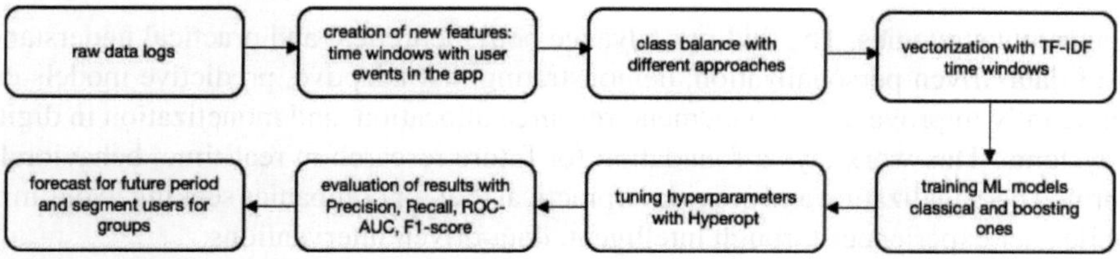

Fig. 2. General methodology for building dynamic segmentation based on target actions.

The methodology integrates multi-scale temporal feature engineering, TF-IDF–based event vectorization, and gradient boosting ensemble modeling, with LightGBM achieving the strongest performance (ROC-AUC > 0.83, high Recall for the minority payer class). Comparative experiments against Logistic Regression, SVM, Random Forest, and other boosting algorithms confirm its superior predictive capability. Feature importance analysis highlights that medium- and long-term behavioral signals—such as repeated payment page visits, historical payment completions, and responsiveness to reminders—are key predictors of future payments. These results support Proposition 1: user behavior in mobile applications exhibits consistent, predictive patterns that enable short-term action forecasting.

The framework operationalizes probabilistic outputs through segmentation thresholds, dividing users into high-, medium-, and low-probability cohorts to guide differentiated engagement strategies. Lift analysis demonstrates that this approach significantly outperforms both random targeting and static, rule-based segmentation schemes, which struggle to distinguish future engagement patterns due to behavioral homogeneity. These findings support Proposition 2: dynamic, model-driven segmentation achieves higher predictive accuracy and adaptability compared to static approaches.

From a practical standpoint, the framework offers flexible forecast horizons, enabling proactive user engagement and resource allocation. Probabilistic segmentation aligns marketing or retention strategies with expected user value, driving conversion optimization. Moreover, the methodology is broadly applicable to domains such as fintech, govtech, and service automation, where fine-grained behavioral modeling is essential.

Nonetheless, several limitations remain. The current design emphasizes dynamic behavioral features while excluding static or quasi-static characteristics (e.g., demographics, aggregate transaction history), which may enhance long-term predictive stability. Future work should investigate hybrid models that integrate dynamic and static features and apply causal inference techniques to uncover actionable behavioral drivers. Additionally, periodic retraining is required to address behavioral drift, underscoring the importance of automated retraining and monitoring pipelines. Finally, segmentation thresholds were empirically defined; future research may employ cost-sensitive optimization or reinforcement learning to refine decision boundaries and improve operational efficiency.

In conclusion, this study contributes a robust, scalable, and interpretable approach to dynamic user segmentation in mobile applications. By combining ensemble machine learning, temporal feature engineering, and probabilistic segmentation, the framework not only enhances predictive accuracy but also provides actionable insights for targeted engagement strategies. The findings advance both theoretical and practical understanding of data-driven personalization, demonstrating that adaptive, predictive models can significantly improve user engagement, resource allocation, and monetization in digital ecosystems. This work lays a foundation for future research in real-time, behaviorally informed personalization and provides a practical tool for companies seeking to optimize mobile user experiences through intelligent, data-driven interventions.

References

1. Duarte, F.: Amount of data created daily. Exploding Topics (2025). https://explodingtopics.com/blog/data-generated-per-day
2. Daoud, J.I.: Multicollinearity and regression analysis. J. Phys. Conf. Ser. **949**, 012009 (2017). https://doi.org/10.1088/1742-6596/949/1/012009
3. Breiman, L.: Random forests. Mach. Learn. **45**, 5–32 (2001). https://doi.org/10.1023/A:1010933404324
4. Pielot, M., De Oliveira, R., Kwak, H., Oliver, N.: Didn't you see my message? In: Proc. of the SIGCHI Conference on Human Factors in Computing Systems (CHI'14), pp. 3319–3328 (2014). https://doi.org/10.1145/2556288.2556973
5. Li, Y., Chu, X., Tian, D., Feng, J., Mu, W.: Customer segmentation using K-means clustering and the adaptive particle swarm optimization algorithm. Appl. Soft Comput. **113**, 107924 (2021). https://doi.org/10.1016/j.asoc.2021.107924
6. Lee, D., Kim, J.: Development of HTC-DBSCAN: a hierarchical trajectory clustering algorithm with automated parameter tuning. Appl. Sci. **14**(23), 10995 (2024). https://doi.org/10.3390/app142310995
7. Apicella, A., Giugliano, S., Isgrò, F., Prevete, R.: Exploiting auto-encoders and segmentation methods for middle-level explanations of image classification systems. Knowl.-Based Syst. **255**, 109725 (2022). https://doi.org/10.1016/j.knosys.2022.109725
8. Cahyani, D.E., Patasik, I.: Performance comparison of TF-IDF and Word2Vec models for emotion text classification. ResearchGate (2021). https://www.researchgate.net/publication/355376912
9. Brownlee, J.: Undersampling algorithms for imbalanced classification. Machine Learning Mastery (2020). https://machinelearningmastery.com/undersampling-algorithms-for-imbalanced-classification/
10. Chawla, N.V., Bowyer, K.W., Hall, L.O., Kegelmeyer, W.P.: SMOTE: synthetic minority over-sampling technique. J. Artif. Intell. Res. **16**, 321–357 (2002)
11. Kovács, G.: A comparison of undersampling, oversampling and SMOTE methods for dealing with imbalanced classification in educational data mining. J. Educ. Comput. Res. (2023)
12. Huang, J., Ling, C.X.: Constructing new and better evaluation measures for machine learning. In: Sangal, R., Mehta, H., Bagga, R.K. (eds.) Proc. of the 20th International Joint Conference on Artificial Intelligence (IJCAI 2007), pp. 859–864. Morgan Kaufmann Publishers Inc., San Francisco (2007)
13. Which metrics are used to evaluate a binary classification model's performance? PI.EXCHANGE (n.d.). https://www.pi.exchange/knowledgehub/metrics-to-consider-when-evaluating-a-binary-classification-models-performance
14. Peduzzi, P., Concato, J., Kemper, E., Holford, T.R., Feinstein, A.R.: A simulation study of the number of events per variable in logistic regression analysis. J. Clin. Epidemiol. **49**(12), 1373–1379 (1996). https://doi.org/10.1016/S0895-4356(96)00236-3
15. Al-Sayed, M.M., Al-Harbi, M., Al-Saggaf, A.A.: Machine learning approach to classification of online users by exploiting information seeking behavior. ResearchGate (2024). https://www.researchgate.net/publication/379501389

16. Breiman, L.: Random forests. Mach. Learn. **45**(1), 5–32 (2001). https://doi.org/10.1023/A:1010933404324
17. Natekin, A., Knoll, A.: Gradient boosting machines, a tutorial. Front. Neurorobot. **7**, 21 (2013). https://doi.org/10.3389/fnbot.2013.00021
18. Chawla, N.V., Bowyer, K.W., Hall, L.O., Kegelmeyer, W.P.: SMOTE: synthetic minority over-sampling technique. J. Artif. Intell. Res. **16**, 321–357. (2002). https://doi.org/10.1613/jair.953
19. Sharma, R.: Data-driven personalization: Revolutionizing user experience. ResearchGate (2024). https://www.researchgate.net/publication/385400466

Author Index

I. K. Koychev (Ed.): ERIS 2025, PINNS 4, p. 107, 2026.
https://doi.org/10.1007/978-3-032-30153-6

MIX
Papier aus verantwortungsvollen Quellen
Paper from responsible sources
FSC® C105338

If you have any concerns about our products,
you can contact us on
ProductSafety@springernature.com

In case Publisher is established outside the EU,
the EU authorized representative is:
Springer Nature Customer Service Center GmbH
Europaplatz 3, 69115 Heidelberg, Germany

Printed by Libri Plureos GmbH
in Hamburg, Germany